Eve E. Hinman Eng. Sc. D, P.E. & David J. Hammond, s.e.

Lessons From the Oklahoma City Bombing

Defensive Design Techniques

Published by

ASCE PRESS

American Society of Civil Engineers
345 East 47th Street
New York, New York 10017-2398

Abstract:
This book documents the Oklahoma City bombing incident from a structural standpoint. It provides detailed discussion of the construction of the building and the hazard mitigation stops taken during the rescue/recovery process. Both authors were present during this process: Dr. Hinman conducted a bomb damage assessment survey to verify blast resistant design concepts that are currently employed, and Mr. Hammond served as Chief Structural Engineer after the incident. This book not only discusses the structural aspects of this particular case, but will also provide insight for all structural and architectural engineers who want to improve the blast resistance of new and existing buildings.

Library of Congress Cataloging-in-Publication Data

Hinman, Eve E.
Lessons from the Oklahoma City bombing : defensive design techniques / Eve E. Hinman and David J. Hammond.
p.   cm.
ISBN 0-7844-0217-5
1. Building, Bombproof. 2. Oklahoma City Federal Building Bombing, Oklahoma City, Okla., 1995. 3. Buildings--Blast effects. I. Hammond, David J.   II. Title.
TH1097.H56 1996   96-49992
693.8'54--dc21   CIP

The material presented in this publication has been prepared in accordance with generally recognized engineering principles and practices, and is for general information only. This information should not be used without first securing competent advice with respect to its suitability for any general or specific application.

The contents of this publication are not intended to be and should not be construed to be a standard of the American Society of Civil Engineers (ASCE) and are not intended for use as a reference in purchase specifications, contracts, regulations, statutes, or any other legal document.

No reference made in this publication to any specific method, product, process or service constitutes or implies an endorsement, recommendation, or warranty thereof by ASCE.

ASCE makes no representation or warranty of any kind, whether express or implied, concerning the accuracy, completeness, suitability, or utility of any information, apparatus, product, or process discussed in this publication, and assumes no liability therefore.

Anyone utilizing this information assumes all liability arising from such use, including but not limited to infringement of any patent or patents.

DEDICATION

This book is dedicated to the hundreds of rescue workers who were in Oklahoma City and the engineers who supported them. With their toil and acts of love, they focused on the individual victims and did their best to overcome the effects of a horrific act.

CONTENTS

ILLUSTRATIONS

PREFACE

My first job after getting my graduate degree was with a defense contractor designing nuclear missile silos and military bunkers in Europe. As it happened, the year that I started working at this firm, 1983, was also the year that major bombings occurred at the U.S. embassies in Beirut, Lebanon, and Kuwait. Not long after these incidents, I found myself working on the development of criteria for the structural design of U.S. embassies against the threat of terrorist attack. To my knowledge, this document was the first of its kind. Over the next 10 years, the art of defensive design for civil facilities evolved. With each new embassy project that came into the office, a little more was learned and applied until finally the process became routine.

During this time, no one thought that such a threat would reach the United States. It was always a problem that existed elsewhere. It was not until the bombing of the World Trade Center in New York City in 1993 that anyone thought that it could happen here. Even after this incident, the rationalization that people held was that New York City is a cosmopolitan city unlike anywhere else in the United States. It bears more resemblance to other international cities like Paris and Rome than San Francisco or Chicago. The reality did not set in until the bombing of the Alfred P. Murrah Federal Office Building in Oklahoma City on April 19, 1995.

David Hammond and I had the truly invaluable opportunity of being present at the site of the Murrah building during the rescue/recovery effort. David's contribution was far greater than mine. He was present at the site for the first 12 days of the effort, on behalf of the Federal Emergency Management Agency (FEMA) Office of Emergency Services, providing expert technical assistance to the effort. The objective of this work was to prevent additional casualties caused by further building collapse and falling debris. During this critical early phase of the effort, he documented in detail all of the engineering issues that arose. Often critical engineering decisions regarding the structural stability of components had to be made within minutes of the issue being identified. As a co-author, David's hands-on experience in disaster situations provides a balance to my own design-oriented approach to the structural response of buildings to explosions.

I gained access to the site of the Murrah building through my boss, Dr. John Osteraas, who took over as Chief Structural Engineer after David Hammond was relieved. Free of any particular task associated with the rescue effort, I had the time to visually document the structural damage to the building. I was present at the site on the last day of the rescue effort, May 5, 1995, and many of the photographs contained in this book were taken on that day.

To date, an enormous amount of literature is available on the design of military facilities; however, very little is available on the design of civilian facilities for an explosive threat. Technical information about the damage to civilian buildings subject to bombings is often not carefully collected, or if it is, remains unavailable to the public. With this book, both the forensic and the preventative design issues are addressed. The hope is that by making this information available to the public, there will be a better understanding of the mechanics of structural failure because of explosions and what can be done to reduce their consequences. Ultimately, it is hoped that by employing defensive design techniques in the design of new buildings, the terrorist will become discouraged from implementing this particularly onerous form of attack.

I would like to acknowledge Paul Weidlinger who taught me everything I know about defensive design. I would also like to acknowledge John Osteraas, whose insight into how buildings fail was invaluable in writing this book.

EVE E. HINMAN

INTRODUCTION

At 9:02 AM on April 19, 1995, the Alfred P. Murrah federal office building in Oklahoma City, Oklahoma, was bombed. The weapon, consisting of a homemade bomb placed in the back of a rental truck, caused the partial collapse of the Murrah building and structural damage to numerous other buildings in the vicinity. The explosion caused the death of 167 persons, including 19 children, and injured 782 persons. Property losses totaled $652 million. Less quantifiable losses include the long-term effects of trauma to the local residents, a significant percentage of whom lost friends and family. Finally, this bombing had a significant impact on the psyche of the American people, who up until a few years ago, had assumed that terrorism was a problem only in other parts of the world.

After the attack, numerous federal agencies assisted the Oklahoma City Fire Department in the rescue and recovery effort, working around the clock for 16 days after the incident. Among the agencies represented on the site were: the Federal Bureau of Investigation (FBI); the U.S. Army Corps of Engineers (USCOE); the Treasury Bureau of Alcohol, Tobacco, and Firearms (ATF); the Federal Emergency Management Agency (FEMA); and Urban Search and Rescue (US&R) teams from around the country. Because of the precarious condition of the remaining structure, in which well over 100 victims were still entombed, engineering played a significant role in identifying and creatively minimizing collapse and falling hazards, without severely limiting rescue operations.

Given that most of the deaths in the Murrah building were caused by falling structural debris, it is fair to say that every floor slab that remained in place was responsible for saving lives. For this reason alone, careful study of this incident from a structural standpoint is warranted.

This book serves as documentation of the incident from a structural standpoint and as a tool for the building owner, security specialist, engineer, or architect to make informed decisions about how to mitigate the effects of an explosion for new or existing buildings. The focus is on protecting office buildings against external vehicle bombs, but the principles are valid for other threats and types of construction as well. The Oklahoma City incident is discussed in Part I. Topics covered include: a description of the weapon used in Oklahoma City, the damage incurred to the Murrah building, and vulnerabilities inherent in the building design. Part II describes the newly implemented US&R system, the engineer's role in that system and this disaster, and how the federal response enhanced the efforts of the Oklahoma City Fire Department and other agencies. Topics covered include: initial hazard assessment, discussion of mitigation, a chronological log of engineering activities, and a guide to building collapse evaluation. Part III provides the designer with practical methods for deterring explosive threats and mitigating damage caused by explosive threats for both new and existing buildings. The emphasis of this section is on cost-effective measures that maintain the architectural integrity of facilities, while providing a higher level of protection to the occupants in case of a catastrophic event.

The authors, both of whom were on-site during the rescue/recovery process, offer firsthand knowledge of the damage and the activities at the Murrah building after the bombing incident. Eve Hinman, who has over 12 years of experience in the design of civilian buildings to mitigate the effects of large-scale vehicle bomb threats, gained access to the site in the latter part of the effort to perform a structural damage assessment. David Hammond, who is the author of Part II, on the rescue and recovery effort, acted as chief structural engineer during the first 12 days of the effort. Mr. Hammond was instrumental in integrating structural engineering into the Urban Search and Rescue system.

Hindsight is 20-20, and the views offered here are with the understanding that the architects and engineers who designed the Alfred P. Murrah building could not have foreseen its tragic end.

Part I. The Oklahoma City Bombing Incident

Chapter 1

The Bomb Used

The weapon used at the Murrah building was a homemade bomb containing an estimated 2,177 kg (4,800 lb) of ammonium nitrate and fuel oil (ANFO). One of the key ingredients, ammonium nitrate, is available commercially as a fertilizer in the form of pellets, typically in 22.7 kg (50 lb) bags.

ANFO is a common explosive used in 95% of the commercial explosives work performed in the United States (Gorman 1995). It is a relatively safe substance that is routinely transported and stored for agricultural use. Even when mixed with an oxidizer such as fuel oil to increase its sensitivity, ANFO is still a relatively stable substance, routinely transported to construction sites in heavy-duty trucks that resemble cement mixers. To make ANFO into a weapon, a booster of high-grade explosive such as pentolite is needed to initiate the detonation.

Despite its relative stability, there are several well-known disasters that have involved the accidental detonation of ammonium nitrate. One incident occurred at a BASF plant in Oppau, Germany, on September 21, 1921. This explosion, consisting of 408,163 kg (900,000 lb) of ammonium nitrate sulphate double salt, created a crater 60 m (200 ft) deep and 130 m (426 ft) in diameter, killing 430 persons and causing severe damage for 6 km. Little is known about how the explosion occurred. It was generally believed, until the time of this explosion, that ammonium nitrate alone was incapable of exploding. The double salt composition manufactured at this plant was believed to be particularly inert and was routinely blasted with dynamite when it caked during storage. The postincident investigation revealed that 45,500–91,000 kg (50–100 tons) of high explosives deliberately concealed under or near a pile of salts was a likely initiator of the event (King and Bauer 1977).

Another well-known explosion occurred in Texas City, Texas, on April 16 and 17, 1947. In this incident, the explosion of 410,000 kg (4,500 tons) of ammonium nitrate in the holding tanks of two docked ships was initiated when one of the ships caught fire. A total of 516 persons were killed, and all of the houses in a 1.6 km (1 mi) radius were destroyed. The damages were totaled to be $67 million (1947 dollars).

The details of how the weapon used in the Oklahoma City bombing was constructed are not known for certain. It is likely that the weapon was fabricated using a combination of boosters, detonators (blasting caps), detonation cord, fusing systems, and plastic containers ("Deposition" 1995). One possible design consists of placing ground ANFO in 55 gal. plastic drum containers. It is possible that detonation cord was used to propagate the detonation from one drum to the next and to ensure effective detonation throughout the volume of the drum. Although explosive devices such as blasting caps and detonation cord are carefully regulated and can not be obtained without a permit, the ATF reports that a small number of these devices are stolen annually (*Arson* 1994).

Grinding of ammonium nitrate is a technique that is known to have been used by terrorists to increase ammonium nitrate's sensitivity to detonation. This is a particularly common technique for ammonium nitrate diluted with inert substances such as calcium carbonate. Members of the European Economic Community (EEC) have mandated the use of diluents, such as calcium carbonate in ammonium nitrate, since 1980 to increase its stability in the event of improper storage ("Council" 1980). However, the United States has no such requirement at the present time. A 1968 patent put forth by Samuel Porter suggests the use of ammonium phosphates or ammonium sulfates to deaden the detonability of ammonium nitrate (Porter 1968), but this approach has never been accepted by any government entity or the fertilizer industry within the United States. Although these desensitizing agents have been shown to be effective in tests using small quantities (a 2 qt ice cream or food container), recent tests have demonstrated that for larger quantities (in 10 gal. containers) these agents are not effective in rendering ammonium nitrate insensitive to detonation (Abbott et al. 1995).

It has been speculated that there was more than one bomb used at the Murrah building. One theory is based on the recording of a seis-

Fig. 1.1. North Elevation of Murrah Building prior to Bombing

mic instrument at the University of Oklahoma, in Norman, 27 km (17 mi) away. This recording is said to contain two separate explosion events. In a report issued by the United States Geological Survey (USGS), the second event was shown to have been caused by different seismic wave phases caused by the single explosion (Holzer et al. 1995). In another report, a retired Air Force general hypothesized that there were demolition charges strapped around four columns of the building in addition to the vehicle bomb outside of the building. The extent of the damage witnessed at the site justifies his hypothesis (Partin 1995). To the knowledge of the writers, this theory has not been accepted by any government authority.

Generically, the large-scale vehicle bomb threat may be defined as a detonation of a large quantity of explosives contained in a stationary or moving vehicle, external or internal to the targeted facility. One of the first notable attacks of this type was the bombing of the marine barracks building in Beirut, Lebanon, on October 23, 1983, by Islamic fundamentalists, which killed 240 persons. The weapon used in this incident contained an estimated equivalent weight of 545 kg (12,000 lb), TNT. This is the largest truck bomb attack executed to date. Other bombings directed against U.S. posts abroad, executed by Islamic fundamentalists, included the Beirut embassy on April 18, 1983; the Kuwait embassy on December 13, 1983; and the Beirut embassy annex in May 1986.

Bombings directed against British targets have been executed for decades by the Irish Republican Army (IRA). Typically, these incidents cause fewer casualties, because the IRA routinely issues warnings prior to executing attacks. The damage caused is typically characterized by extensive window breakage and business interruption.

Large-scale vehicle bombs made their debut in the United States on February 26, 1993, with the bombing of the World Trade Center in New York City. The weapon used in this incident was constructed of the fertilizer urea nitrate. The size of the weapon, which detonated in the underground parking garage, consisted of approximately 545 kg TNT equivalent (1,200 lb). This explosion killed six persons and injured more than 1,000 persons, mostly because of smoke inhalation. Given the large number of occupants in this complex (roughly 150,000 people work and visit the complex daily), it is remarkable

that so few died or were seriously injured as a result of this incident. Structural damage included a five-story subgrade crater that measured 24–36 m (80–120 ft) across on some levels. It cost over $1 million per day in lost rent alone while the facility was closed for 51 days of structural repairs (Harvey 1995). After this event, the owner, The Port Authority of New York and New Jersey, spent over $300 million in rehabilitation directly related to the damages caused by the explosion. Following this incident, the bombing in Oklahoma City was the next largest vehicle bomb explosion in the United States and the largest vehicle bomb attack in this country, to date.

The subsequent two attacks directed against U.S. targets in Saudi Arabia in the 15 months following the Oklahoma City incident (one in November 1995, killing five persons and the other in June 1996, killing 19 persons), as well as the resurgence of explosive attacks by the IRA and Palestinian terrorist groups in this same time frame, indicate that this threat is not likely to dissipate in the foreseeable future.

CHAPTER 2

THE MURRAH BUILDING DESIGN

In retrospect, there is little that could have been done to structurally prevent extensive damage to the Murrah building. The weapon used was too large and too close to the building. However, it is useful to review the design of the Murrah building with defensive design principles in mind. Such measures may have been effective in mitigating damages to a limited degree and would have undoubtedly helped to reduce damage levels if a smaller weapon had been used.

The Murrah building was a nine-story, reinforced concrete frame structure designed in 1974. The rear of the building, shown in Figure 1.1, faced north and had curbside access to vehicles. For convenience, an indented loading zone was provided at the rear entrance of the building, 2.3 m (7.5 ft) from the building exterior.

To optimize the natural light available, the north facade was glazed with a full height window/wall system. A glass curtain wall was an unfortunate choice of cladding for this face, given its proximity to the street. Glass is highly vulnerable to a blast, because it breaks at low pressures, creating hazardous shards.

The Murrah building was surrounded by buildings having a broad range of functions: commercial, governmental, retail, residential, and religious. Directly across from the Murrah building, on the north side, were the Journal Record building (behind a large parking lot), the Oklahoma Water Resources Board building, and the Athenian Restaurant building. To the northeast was the YMCA, and to the northwest was the Regency Tower apartment complex. On the south side was the Federal Courthouse. Churches were located on the other two sides: The First Methodist Church on the east side and St. Joseph's Cathedral on the west side.

The walls of the stairwells and elevator shafts for the building project from the south side of the building. These shear walls provided the principal lateral bracing for resisting the explosive forces of the weapon. To protect the south face from direct sunlight, precast concrete shades were used (see Figure 1.2). On the east and west facades there were 7.62 mm (3 in.) thick granite infill panels (visible in Figure 1.1).

The front of the building faced south onto a large plaza more than 46 m (150 ft) in depth (see Figure 1.3). Having a large setback from the street is recommended for protection against vehicle bombs, but the accessibility of the north face eliminated the benefit that could have been derived from this feature. By evenly distributing the setback between the north and south sides, a more balanced design would have been achieved with more inherent protection.

Other distinctive architectural features of the building were the four large cylindrical concrete air ducts at each of the four corners of the building (the two on the west face are visible in Figure 1.1). Also, two one-story annex structures were on the east and west faces. The west annex structure, which functioned as the loading dock area for the building, was relatively undamaged after the attack and was used as the command post for the rescue effort.

Beneath this plaza was a four-level underground parking garage. Having an underground garage adjacent, rather than directly beneath a building, is generally encouraged in blast-resistant design to mitigate damage to the building in the event of an explosion in the garage. Unfortunately, it did not help in this design, because the garage was not as attractive a target as the easily accessible north side.

The building housed a broad range of federal agencies. The functional layout by floor was:

1st Floor   Social Security Administration
             General Services Administration
2nd Floor   Child Care Center

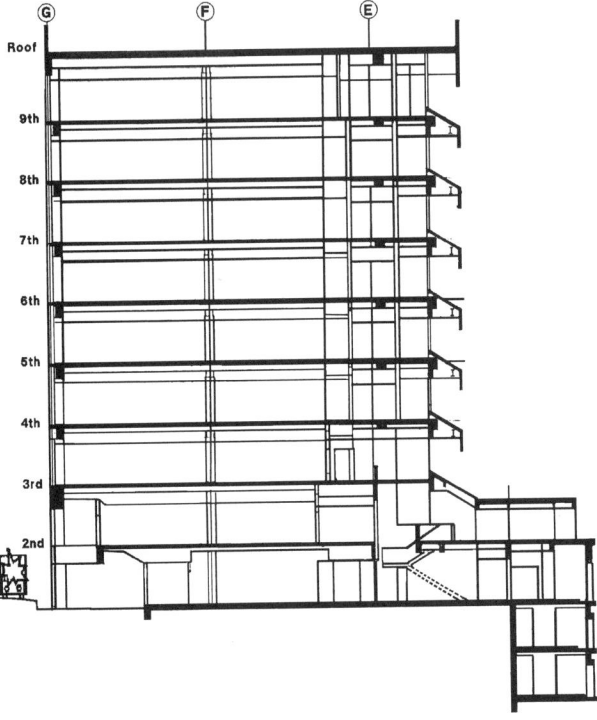

Fig. 1.2. Section through Building Width

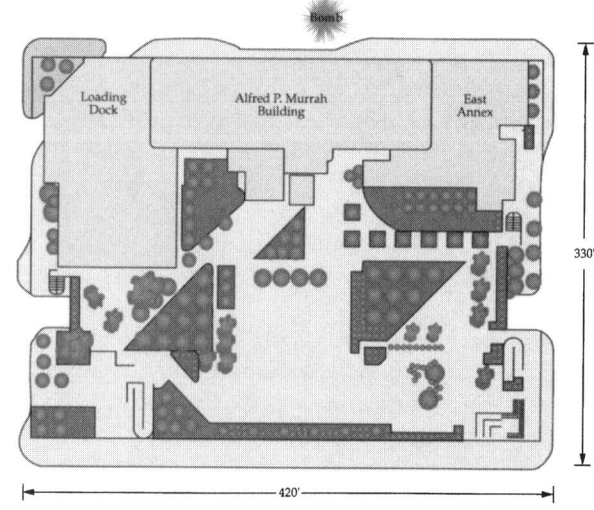

Fig. 1.3. Site Plan of Alfred P. Murrah Federal Office Building

3rd Floor   Defense Audit Agency
            Federal Employees Credit Union
            Housing and Human Services
            U.S. Army
            General Accounting Office
4th Floor   Federal Highway Administration
            Army Reserve
            Army Recruiting
            Snack Bar
5th Floor   Department of Agriculture
            Housing and Urban Develop-
               ment (HUD)
            U.S. Customs
            Veterans Administration
6th Floor   U.S. Marine Corps
            Vacant office space
7th Floor   HUD
            Drug Enforcement Administra-
               tion (DEA)
8th Floor   HUD
9th Floor   Secret Service
            ATF
            DEA

The floor plan of the Murrah building tower, shown in Figure 1.4, measured 61 m × 23 m (200 ft × 75 ft). In layout, the building was divided into 10 bays in length and two bays in width with three lines of columns across the width (E, F, and G). The typical floor to floor height was 4 m (13 ft). Floor slabs, 15.24 cm (6 in.) thick spanning 20 ft in the east-west direction, were supported by beams, 18.9 cm × 7.8 cm (48 in. × 20 in.) spanning 10.7 m (35 ft) in the north–south direction. The beams were dimensioned so that only nominal shear reinforcement was necessary. Additional shear reinforcement along the entire length of the beams may have improved their performance by providing confinement and promoting a ductile response. Also, by designing the building to be only two bays wide, limited redundancy was provided to

the structure. Generally, the outermost bay is most vulnerable to collapse in large explosive events. For this building, the outer bay comprised a full one-half of the width of the tower.

The first floor plan, shown in Figure 1.5, indicates that the north elevation (column line G) was supported by four freestanding columns, each measuring 14.2 cm × 7.8 cm (36 in. × 20 in.) at the base. These columns spanned the first two floors without lateral support in the east-west direction. At the third floor level, there were 12.2 m (40 ft) long transfer girders, measuring 1.52 m × 0.91 m (5 ft × 3 ft), which supported intermediate columns (this is shown in Figure 1.1). Architecturally, this design served to provide an attractive, open entranceway into the building. However, it reduced the redundancy of the design so that the loss of a single transfer girder would cause the immediate collapse of a footprint area measuring 12.2 m × 10.7 m (40 ft × 35 ft). Similarly, the loss of a single primary column would cause the collapse of a footprint area measuring 24.4 m × 10.7 m (80 ft × 35 ft). These losses do not include the additional bays lost from progressive collapse.

The building was designed primarily for gravity loads. Lateral loads consisted of 128.72 km/h (80 mph) wind forces and Uniform Building Code zone 1 seismic requirements. The material properties used were:

28 day compressive strength of concrete: 27.5 MPa (4,000 psi)
Yield strength of the reinforcing bars: 413.7 MPa (60,000 psi)

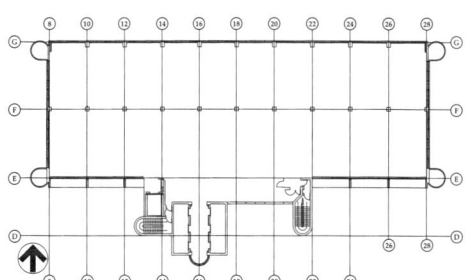

Fig. 1.4. Typical Floor Framing Plan (Floors 4–8)

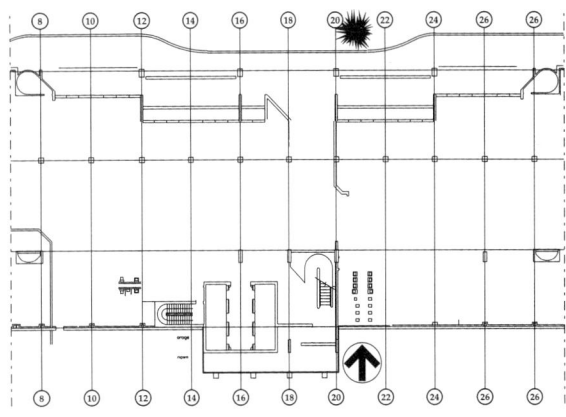

Fig. 1.5. First Floor Framing and Foundation Plan

It appears that some type of small explosive, such as a pipe bomb, may have been at least discussed during the design of this facility to counter the threat of Vietnam War protesters attacking the facility. Based on the information provided on the design drawings, it is unlikely that structural solutions were considered for thwarting this type of attack. It is more likely that architectural and security solutions were used. Architectural measures may have included placing vital functions of the building away from vulnerable areas such as the lobby.

CHAPTER 3

BUILDING DAMAGE

The explosion occurred within a truck parked in the passenger loading zone outside if the north entrance, nearest to column G20. The center of gravity of the weapon was roughly 3.05 m (10 ft) away from the building. The explosion generated a shock wave that propagated outward, in all directions, from the source. Because of the irregular shape of the weapon and its nonideal design, the shape of the shock wave was not perfectly hemispherical. (Although for convenience, it is approximated to be hemispherical in this discussion.)

Weapons Effects

Figure 1.6 shows the rapid decay of the incident pressure with distance from the explosion. The north face of the building, which was in the direct path of the shock wave, was subject to reflected pressure levels. These can be significantly higher than the pressures shown in Figure 1.6. For example, the normal reflected pressure 10 ft from this weapon is 130 MPa (18,810 psi). As the shock wave engulfed the building, the loading became progressively more complex because of diffraction of the shock wave around corners and reflection from the building surfaces. Note that the incident pressures shown in Figure 1.6 are several orders of magnitude higher than the pressure for which conventional buildings are designed (hurricane winds may have a peak pressure less than 1 psi). Unlike conventional loads, however, these air-blast loads decay exponentially to zero over a duration of several milliseconds. In addition to the air-blast pressures acting on the building, a portion of the energy was imparted into the ground, creating a crater 9.1 m (30 ft) wide and 2.4 m (8 ft) deep. (A more complete description of weapons effects associated with the explosion is given in Part III.)

Overall Structural Damage

Damages observed may be categorized as direct air-blast damage and collateral effects caused by progressive collapse. Progressive collapse may have been initiated by impacting debris and destabilization of the remaining components. The worst damage was observed on the north side of the structure, which was closest to the explosion and subject to reflected pressures. Figure 1.7 (see Color Section) shows the structural debris piled 30 ft high along column line F. It is clear from this figure that the force of the explosion was many times greater than the capacity of the members. Reinforcing bars, fragments of slabs, remnants of false ceilings, and ductwork appear to be dripping off the floor levels like confetti and ribbon. On the east face (see Figure 1.8), granite infill panels were lost, particularly at the lower floors, where the pressure levels were greatest. Both the inside and outside surfaces of the exterior walls received significant pressure loadings. The west side (see Figure 1.9) was farther from the source and was found to be in relatively good condition after the explosion. The front face (see Figure 1.10) also was in good structural condition, except that the windows had been broken and precast connections had been broken in a number of locations.

This explosion caused the collapse of nearly

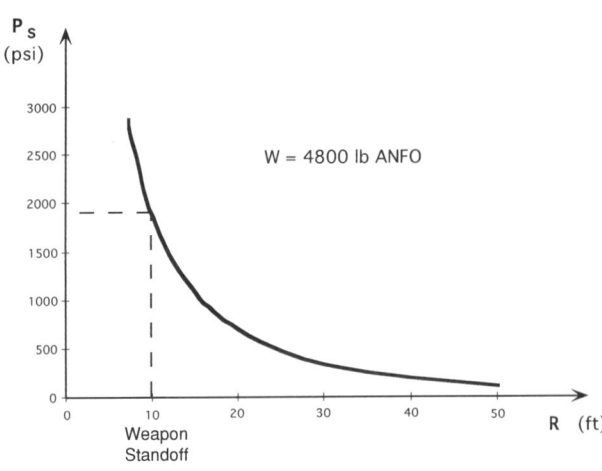

Fig. 1.6. Peak Incident Pressure versus Range

one-half of the building. Figure 1.11 shows the outline of the remaining structure. Three of the four primary columns supporting the transfer girders collapsed (G16, G20, and G24), bringing down eight of the 10 bays along the north side of the building. Also, one interior column (F24) collapsed, bringing with it the two adjoining bays and creating a "bite" in the outline of the remaining structure. Although interior columns F20 and F22 remained standing, floor slabs attached to these columns were lost on the second and third floors, creating unbraced column lengths that spanned three stories. A floor plan view of this floor slab damage is shown in Figure 1.12.

Sequence of Collapse

Based on an understanding of the general behavior of structures subject to explosive loads, first principles calculations, and observed dam-

age at the site, it is concluded that direct air-blast effects caused the failure of column G20 and floor slabs on the lower levels, which in turn initiated the global collapse of columns G16, G24, and interior column F24.

Column G20, closest to the explosion, failed due to brisance or the shattering of concrete. Loss of structural integrity due to this damage mechanism occurs for explosions that are close-in to structural components. It is caused by multiple reflections of the shock wave within the material. The shock wave, which is initially compressive propagates through the concrete, reflects from the opposite side of a structural member as a tensile wave. Such reflections may cause spalling and microcracking. Confinement, provided by means of helical reinforcement, closely spaced ties, or steel plated caisson designs are methods for mitigating this failure mode.

It is unlikely that concrete shattering was responsible for the collapse of columns G16 and G24, which are considerably further away from the explosion than G20. The condition of these columns after the incident indicates that they failed between the second and third floors in response to the transfer girder rotating inward.

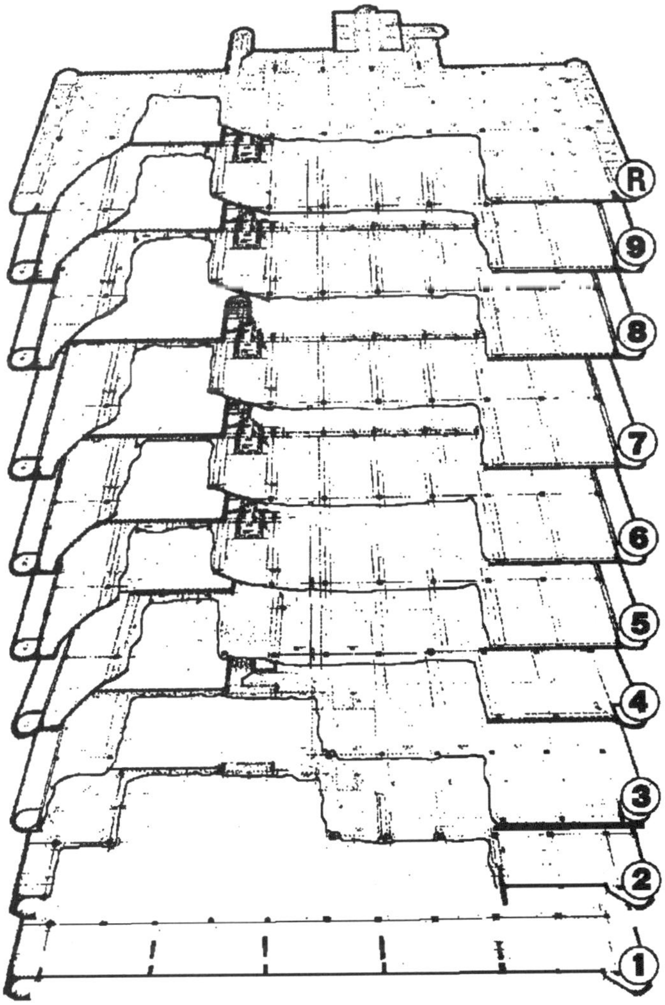

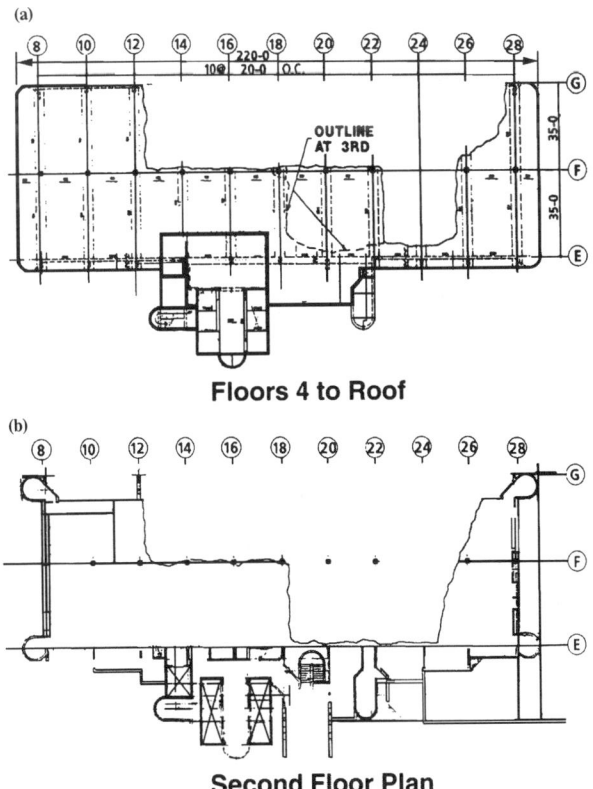

Fig. 1.11. Outline of Remaining Structure

Fig. 1.12. Postexplosion Floor Plans

Evidence of concrete shattering and microcracking was witnessed during debris removal. Apparently, intact elements crumbled when lifted (see Figures 13a and b), a behavior attributed to microcracking. Extensive shattering of concrete was the cause for the small void spaces in the debris pile, worsening the chances of survival for the trapped occupants. Though this effect was in part due to the close-in effects of the explosion, another factor may have been the impact of the debris when it came crashing to the ground.

Calculations indicate that floor slab failure due to direct air-blast effects is expected for the second through fourth floors, between column lines 16, 24, F, and G. On the lower floors where the air-blast had relatively unimpeded access to the building interior, the slab damage extended back to the south face of the building along column line E between lines 18 and 26 on the second and third floors, consistent with calculated damage patterns. Slab failure in conjunction with the loss of column G20 then caused the buckling of the exterior frame along the G line at the lower floors. Lateral propagation of the failure to neighboring bays was caused by destabilization of these members and/or impact by their neighbors. This collateral effect may have been exacerbated by prior weakening of these elements by direct air-blast effects.

Floor slabs tend to be vulnerable to explosive effects due to their large surface area compared with other elements, such as columns. [This viewpoint is supported by the U.S. Corps of Engineers (Conrath and Walton 1995).] Floor slabs are first subject to an upward and then a downward loading from the explosive forces. The upward loading causes cracking and weakening of the members in shear. This is followed by a downward pressure, causing collapse. As the slab deflects downward, it pulls the outer frame inward and causes this already weakened system to fail by buckling at the lower floors. Shear failure rather than flexural or bending failure occurs due to the extremely high intensity and short duration of the loading close—in to the explosion. In short, the slab is sheared off before it has a chance to bend.

On the lower floors where the air blast had a relatively unimpeded access to the building interior, the slab damage extended back to the south face of the building along column line E between column lines 18 and 26 on the second and third floors.

In addition to the failure of the outer frame between column lines 12 and 28, one inner column was lost. This column, F24, was not as close to the explosion as columns F22 or F20, which remained standing. There are several possible explanations why column F24 collapsed rather than column F20 or F22. All three columns were close to collapse due to loss of lateral support at the second and third floor levels. Column F24 may have been subject to impact by falling debris which triggered its failure, whereas its neighbors were spared this additional assault. Another explanation is that the lateral support provided by the elevator core and stairways helped to keep columns F20 and F22 standing, whereas column F24 did not have the benefit of this additional resistance.

Observed Failure Modes

Damage patterns along the interior columns line "F" ranged from total collapse to undamaged. Figure 1.14a shows a two-story length of failed column F24 (spanning from the first to the third floor level) found intact within the debris. Figure 1.14b is a close-up of this column at the second floor level. Little evidence of the connecting beams is present, indicating nonductile failure at the connections. Figure 1.15 shows the two neighboring columns, F20 and F22, which remained standing but were unsupported for three-stories heights (except for debris piled up on the outside face). To provide additional stability to these columns during the rescue effort, grout was placed at the floor level connection points, and steel tubes were used to brace the columns (see Figure 1.16). Similar damage was incurred in the World Trade Center explosion, in which there was extensive floor slab damage, causing unbraced lengths of columns to require stabilization immediately after the incident. Floor slab failure is common in explosive events because of the large surface area upon which the air-blast can act and the fragility of these thin structural components.

Continuing to move farther away from the explosion source, the beam behind column F18 was badly damaged by direct air-blast effects (see Figure 1.17a). The bottom reinforcement is acted as a catenary with the concrete portion of the beam resting on top. Paper caught beneath the concrete is evidence that the fragment first lifted up before falling into its present position. A closer look reveals that there was a cold joint near the midspan of the beam (see Figure 1.17b).

Figure 1.18 shows column F16 intact with the failed supporting beam behind it. The slab-column connection shows signs of classical punching shear failure with segments of the beams still attached to the column. This failure mode is somewhat less severe than for columns F20 and F22, which were gouged out at the column-beam connection due to direct shear failure. For columns to the west of this point, the column-beam connections were intact.

Figure 1.19 shows the remaining portion of the transfer girder adjacent to column F12. It shows that the girder failed at midspan where an intermediate column had been supported. The brittle shearing of column F14 through the transfer girder at this location was caused by progressive collapse initiated by the loss of its neighbors to the east. A closer look at this member in Figure 1.20 reveals that the bottom reinforcement had been ripped out by the failed portion of the girder, pulling straight the shear ties in the members. Behind the transfer girder, Figure 1.21 shows that the floor slab behind the transfer girder had been pulled down with the column. Floor slab failure extends upward to the roof along this column line.

Figure 1.22 shows typical floor slab damage at the edge of the collapsed region at column line 12. Observe in this figure that the length of the remaining slab corresponds to the length of the negative reinforcement provided, indicating that the negative reinforcement participated in main-taining the integrity of the slabs. Also note that the positive reinforcement was broken off at the beam, indicating that the positive reinforcement was not effective in transmitting the load to the supporting beam elements. The pristine condition of the beam compared with the slab further supports the observation that the load was not adequately transferred to the beam.

The failure modes shown in Figures 1.14–1.22 are all brittle, caused by the close-in explosive effects and progressive collapse. More ductile failures resulting from flexure were difficult to find among the debris. One area in which flexure was observed was in the east annex structure. Here, the concrete at the columns was spalled because of the upward flexure of the slab system (see Figure 1.23). This damage was caused by air-blast pressure that became trapped in this unwindowed annex, which acted upward on this slab with relative low-pressure levels, but for a longer period of time than the exposed members in the main portion of the structure.

On the front face of the building, only the precast panels at the third floor level exhibited noticeable damage (see Figure 1.24). These members failed at their connections and had to be supported by using chains. This response is typical of precast members, which tend to have a lower resistance to explosive loads, due in part to the lower capacity of the connections.

Fig. 1.29. The First Methodist Church

In summary, the observed failure modes for the Murrah building were brittle and occurred mostly at the connections. This was because of the excessive forces of the explosion, progressive collapse, and the nonductile construction used in the design of this building. Other typical failure modes included the microcracking of members and the ripping out of reinforcing bars from the members. Only in regions in which the pressures were relatively low was there evidence of a ductile flexural response (i.e., the east annex structure).

Damages to Other Buildings

Damage to other buildings in the vicinity of the Murrah building is shown graphically in Figure 1.25. Most of the serious structural damage extended to the north of the Murrah building. This is because the Murrah building acted as a buffer shielding the south side from explosion effects. A total of 11 buildings collapsed (see Figure 1.25). All of these were constructed of unreinforced masonry, which is known to perform poorly under stress. In addition, 23 buildings sustained severe damage, and 300 buildings sustained some level of damage within a 2.6 sq km (1 sq mi area). Window glass was broken within 3.2 km (2 mi) of the building. Buildings in the vicinity are shown in Figures 1.26–1.32.

Fig. 1.30. St. Joseph's Cathedral

Fig. 1.32. YMCA building

PART II. THE RESCUE/RECOVERY EFFORT

CHAPTER 4

ORIGIN OF US&R TASK FORCE

The devastating 1985 Mexico City earthquake, followed by earthquakes at San Salvador (1986) and in Armenia (1988), clearly demonstrated that successful search and rescue in urban environments required the use of highly trained and especially equipped personnel.

Rescue teams from many countries, including the United States, responded to these disasters and were confronted with complicated, heavy concrete structures that entombed numerous live victims. Successful rescue was dependent on speedy removal of the victims as well as their proper medical treatment (which needed to be initiated prior to release from confinement). The thoughtful coordination of the search, rescue, medical, and technical aspects of Urban Search and Rescue (US&R) was crucial but very difficult to accomplish in this chaotic, disaster environment.

For the most part, separate groups of search specialists (mostly canine), heavy rescue fire-fighting specialists, medical responders, and a few engineers constituted the response to these earthquakes. Their efforts were hampered by lack of coordination and unfamiliarity with each other's location, needs, and capabilities. In a few cases, coordination between technical and canine search followed by rapid deployment of heavy rescue groups led to a dramatic positive removal of entombed victims. However, during the two weeks following the Mexico City earthquake, approximately 150 entombed live victims were removed, but about the same number of rescuers perished.

During 1986, within the United States, a number of search and rescue organizations, including public agencies, began to focus on the concept of creating well-trained and equipped, multidisciplined groups to respond to the problems of rescue in the urban environment (US&R). The dialogue was initially focused between the relatively well-established canine search groups and heavy rescue firefighter entities, but it was augmented by volunteer doctors who had experienced confined space medical

problems in coal mine cave-ins. The few structural engineers that were familiar with the needs also became part of this movement.

Planning at Federal Level

The Robert T. Stanford Disaster Relief and Emergency Assistance Act of 1988 substantially increased the role of the federal government in disaster response and recovery. This act revised and amended the Disaster Relief Act of 1974 to expand the scope of disaster relief programs and defined the role of the government in all four phases of the disaster management cycle: preparation, response, recovery, and mitigation. With the broadening of the federal role in disaster response, it soon became apparent that there was a need to coordinate the efforts of the various federal agencies that had the capabilities to respond to disasters. The Federal Emergency Management Agency (FEMA) was created in 1979 and began a multiagency planning effort that led to the publishing of the *Federal Plan for Response to a Catastrophic Earthquake* in 1987.

This plan represented an agreement among the various federal agencies with disaster response capabilities and the role each would play in a catastrophic earthquake. The plan was oriented primarily towards a major earthquake in California and represented an "all or nothing" approach; the entire plan would be implemented if needed. The plan was tested during a major exercise in California during the summer of 1989. Hurricane Hugo and the Loma Prieta earthquake of 1989 demonstrated the value of preplanning but also pointed out several problems with the plan. In each case, the catastrophic event envisaged by the plan did not occur, yet the capabilities of local governments were severely taxed. Some federal assistance was needed, but the requirement for a massive federal response did not materialize. Federal agencies were unsure how to react to a less than full activation. It was also obvious that the plan would need to focus on all hazards and to include events that were less than catastrophic. The need for a multihazard, flexible plan was further confirmed during Operation Desert Storm, since plans that had been developed for a full mobilization had to be downsized.

On the basis of the lessons learned from the Loma Prieta earthquake and Hurricane Hugo, FEMA coordinated a new planning effort that resulted in the current Federal Response Plan (FRP). This new plan represents a coordinated approach to providing response for a variety of disasters. Its hallmark is flexibility, allowing federal officials to activate portions of the plan appropriate to the level of response required. Under the plan, federal activities are coordinated by an emergency response team (ERT) headed by a presidential-appointed federal coordinating officer (FCO). All federal emergency capabilities and assets are grouped by function rather than agency, allowing for a fast, coordinated response to requests for assistance. A functional unit is called an emergency support function (ESF) and is headed by a lead agency charged with coordinating the activities of that function.

FEMA's US&R Initiative

In June 1989, FEMA's Office of Emergency Management and the U.S. Fire Administration met to review the lessons learned from the Armenia earthquake. The status of U.S. US&R was reviewed, and the possibility of a national US&R initiative was discussed. Several problems became evident:

- Traditional search and rescue was largely confined to a rural environment with only a limited number of search dogs having been trained for the urban environment.
- Local fire departments who were charged with US&R in most jurisdictions, were not equipped for heavy rescue (lacking specialized tools and training) and would be overwhelmed by the fire fighting requirements of a catastrophic event.
- There were no nationally recognized standards for personnel, training, or equipping of US&R teams.

Shortly thereafter, Hurricane Hugo and the Loma Prieta earthquake clearly demonstrated that there was an extreme danger from structural collapse and that such collapses would immediately overwhelm local resources. During the response to these events, it became evident that well-trained and equipped medium and heavy US&R teams did not exist in the United States and that the federal government had no way to identify and mobilize any existing assets.

It was clear that a new and extensive approach was required to meet this need.

In 1990, a number of working groups met under the sponsorship of FEMA. The recommendations of these working groups were incorporated into the base documents that defined the FEMA US&R program. In fiscal year 1990, Congress approved a supplemental appropriation of $800,000 for this US&R planning and approved an additional $1.975 million in fiscal 1991 for the formation of the National Urban Search and Rescue System. In May 1991, FEMA invited state and local jurisdictions to apply as sponsoring jurisdictions for US&R task forces and to be eligible for grants on a 50/50 cash-match basis. In August 1991, a technical review panel screened the 34 applications received, and on the basis of the panel's recommendations, FEMA selected 25 jurisdictions as sponsoring agencies [see illustrations of task force (TF) locations (Figure 2.1) and composition].

US&R Task Force

The basic US&R task force consists of 56 persons divided into four teams: search, rescue, medical, and technical, led by two task force leaders (Figure 2.2). The task force is designed to be self-sufficient for 72 hours and to operate in two, 12 hour shifts. It is expected to be committed for an operational period of 10 days and to be resupplied and supported by the Department of Defense (DOD). Each member of the task force must meet the same basic entry requirements for experience and training and must be

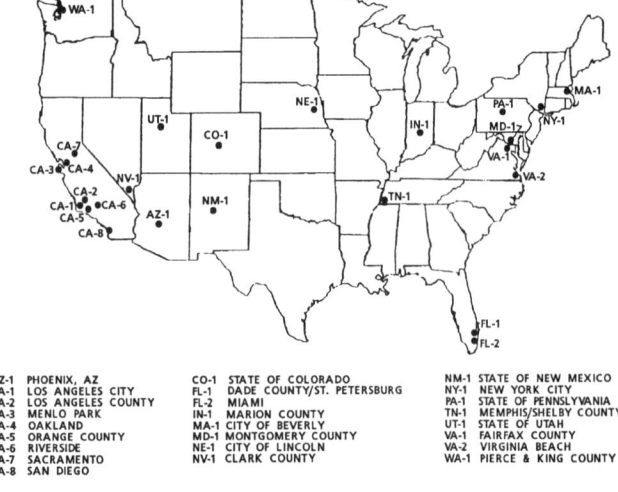

AZ-1	PHOENIX, AZ	CO-1	STATE OF COLORADO	NM-1	STATE OF NEW MEXICO
CA-1	LOS ANGELES CITY	FL-1	DADE COUNTY/ST. PETERSBURG	NY-1	NEW YORK CITY
CA-2	LOS ANGELES COUNTY	FL-2	MIAMI	PA-1	STATE OF PENNSYLVANIA
CA-3	MENLO PARK	IN-1	MARION COUNTY	TN-1	MEMPHIS/SHELBY COUNTY
CA-4	OAKLAND	MA-1	CITY OF BEVERLY	UT-1	STATE OF UTAH
CA-5	ORANGE COUNTY	MD-1	MONTGOMERY COUNTY	VA-1	FAIRFAX COUNTY
CA-6	RIVERSIDE	NE-1	CITY OF LINCOLN	VA-2	VIRGINIA BEACH
CA-7	SACRAMENTO	NV-1	CLARK COUNTY	WA-1	PIERCE & KING COUNTY
CA-8	SAN DIEGO				

Fig. 2.1. FEMA US&R Task Forces (25 Total within U.S.)

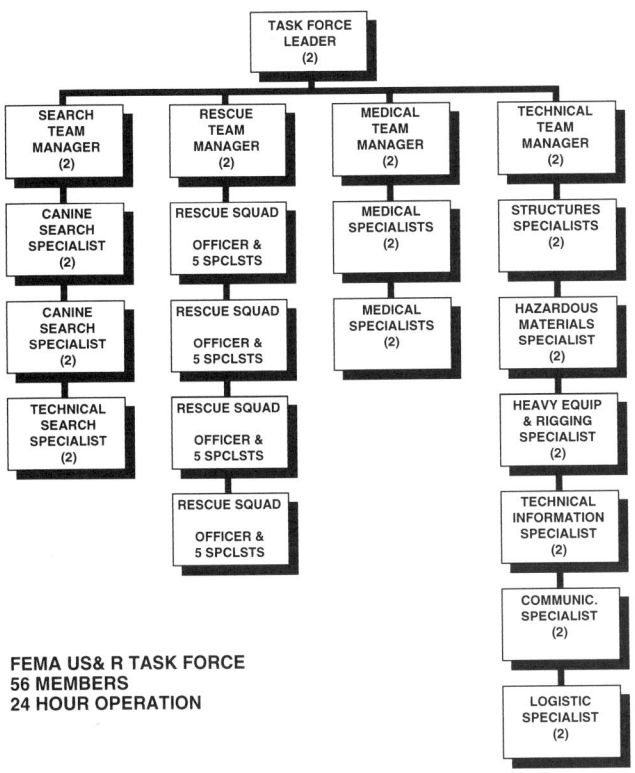

FEMA US&R TASK FORCE
56 MEMBERS
24 HOUR OPERATION

Fig. 2.2 FEMA US&R Task Force

fully-qualified in his or her own specialty. Cross-training is strongly encouraged. When not activated as part of the federal US&R response system, the task force can be used by the sponsoring jurisdiction for local emergency work, deployed either as teams or as a full task force.

The following elements make up each FEMA US&R task force:

1. A search team composed of two team leaders, four search dog teams (a dog handler and search dog), and two technical search specialists. The mission of the team is to locate live victims in collapsed structures using a variety of canine, electronic, and physical search strategies.

2. A rescue team consisting of two team leaders and four rescue squads composed of one officer and five rescue specialists. The team is responsible for evaluation of compromised areas, structural stabilization, breaching and site exploration, and the extrication of live victims.

3. A medical team including two team leaders who must be physicians and four medi-

cal specialists (emergency medical technicians or nurses with emergency medical training or experience). The team's primary purpose is to minimize health risks, treat task force members, and support personnel (including canines). A secondary mission is to stabilize victims prior to extrication and provide treatment until the victim can be turned over to local medical units.

4. A technical team consists of two team leaders and 12 technical specialists. The team is responsible for the evaluation of hazardous or compromised areas, structural assessment, stabilization advice, hazardous materials monitoring, liaison with local capabilities, communications, logistics, and information/documentation. This team includes two structural specialists (structural engineers).

Upon a presidential declaration of disaster, the FRP is activated. This activation may be for the entire plan or only those ESFs that may be needed. If the nature of the disaster is such that US&R capability may be required, FEMA will place selected task forces on alert. Once the need for US&R assets is confirmed, the task forces will be activated and have six hours to move to a designated point of departure. The task force will then be transported to a mobilization center near the site of the disaster. Once committed, operational control of the task force passes from FEMA to the local incident commander. The task force would continue to be supported by FEMA and resupplied by the DOD.

Task force training must take place on various levels. There is a requirement for basic training for each member of a task force, such as first aid, incident command system, and basic rescue. Technical training is also required for each specialist on a task force, and the task force must undergo team training designed to foster cohesiveness and unity of purpose. Unfortunately, in 1991, there were no national standards or existing courses that met these requirements.

To fill this need, FEMA required the working groups in the areas of search, rescue, medical, logistics, communications, and technical to survey existing training resources and identify the shortfalls. When training deficiencies were identified, actions were taken to either develop the needed course by the working groups, or to place a contract for development with an appropriate source.

Beginning in April 1992, FEMA conducted six orientation training sessions across the country. These four-day sessions were designed to prepare a task force to operate as part of the national US&R system and focus on the FRP, agency responsibilities, and task force operations. Participants included task force leaders, team leaders, technical specialists, and DOD liaison officers. Other training courses that have been developed are:

- Crush syndrome/confined space medicine training (FEMA contract)
- Communications training (developed by the Boise Inter-Agency Fire Center)
- Structural specialist training (developed by the U.S. ACE)
- Canine Search Training (developed by working group)
- Logistic specialist course (developed by working group)
- Advanced rescue specialist training (developed by working group)
- Technical rescue training (developed by working group)

Engineers' Participation in Response to Disasters and US&R

In the late 1970s, Disaster Emergency Services (DES) committees were started in California by both the American Society of Civil Engineers (ASCE) and the Structural Engineers Association of California (SEAOC). The focus of these groups was to develop and organize the response of engineers to aid building departments in the postearthquake safety evaluation of damaged structures. Engineers volunteered in this capacity following the Coalinga earthquake in 1984, and this activity has become a vital part of the planned response to earthquakes in California.

The Applied Technology Council publication, ATC-20 "Procedures for Post Earthquake Safety Evaluation of Buildings," funded in 1989 by the Office of Emergency Services, State of California (CAOES) and FEMA, was developed with guidance from members of these DES committees and has become standard methodology. Many engineers throughout the United States have been trained for this type of postearthquake volunteer response.

After the Mexico City earthquake experience, a few engineers perceived the need to work within the rescue, firefighter community to enhance their knowledge and capability in dealing with heavy, complicated structures. Through the National Association for Search and Rescue and a California based nonprofit group, Urban Search and Rescue Inc., discussions and training sessions were conducted in an attempt to add to the knowledge base regarding this type of collapsed structure.

In 1990, these engineers became members of FEMA's Steering Committee and Working Groups, thereby contributing to the development of the national US&R program. Shortly thereafter, USACE sponsored the development of the Structural Specialist Training Course. This course was designed to train both engineers from USACE as well as those civilian engineers who would become task force structural specialists. The training course had been conducted six times by the end of 1995, and 173 engineers have been trained as structural specialists.

Nationwide, engineers volunteered to become part of a task force by responding to an article in the August 1992 issue of *ASCE News*. They were subsequently added to the national database as well as being made known to their nearest task force (more than 300 engineers responded). Within California, the DES Committee of SEAOC was able to obtain responses from about 40 volunteer engineers who were interested in becoming task force participants. In other cases, the individual task forces directly contacted engineers within their government agencies or communities.

CHAPTER 5

ROLE OF THE ENGINEER

Each task force is intended to have six specially trained civil/structural engineers to be able to deploy with two engineers in any six hour period. These volunteer engineers (S. Spec.) are required, in addition to normal US&R training, to attend the one-week special course developed by USACE. This learning experience is intended to familiarize the S. Spec. with the disaster environment. They learn that US&R most often occurs in buildings that are fully or partially collapsed and very dangerous. These buildings will normally be multistoried structures containing heavy debris with a high potential for additional collapse. US&R teams with specialists trained in search, rescue, and medical care will need to work in this dangerous environment when there is a probability of recovery of live victims. Engineers trained and experienced in damaged building evaluation can help reduce (or at least better define) the risk to these teams and the victims. In order to function effectively, these engineers must also be well prepared to make difficult value judgments in an environment that is very different from the orderly design office. By contrast, the search, rescue, and medical members of the team are asked to make rapid, high-pressure decisions as a normal part of their occupations.

The normal firefighter's tendency will be to take a significant risk to save a life. In the absence of any engineering advice, or before it was available, many victims have been saved in burning structures, explosions, earthquakes, etc. Engineers must understand that even though they will feel "responsible" for operations that take place in undesirable conditions, they need to give their best possible assessment and advice, including their degree of uncertainty. The leadership (TF leader and rescue team leaders) will consider the engineer's input, along with others, make *their* decision, and proceed (or not) with the operation while attempting to minimize the risk. *The engineer's role is to provide the task force with critical information, not to make all the critical decisions.* They will, however, have to accept the fact that there will often be a conflict in primary focus. (Rescuers focus on *saving victims*, engineers focus on *rescue safety*.) As with most engineering problems, the first job will be to identify the problem. Some of the basic questions to be addressed are:

- What types of structures are involved
- What hazards are present—falling, collapse, other
- What are the locations and conditions of remaining voids in the structure
- What are the locations of previous access openings in the structure
- What tools and shoring/stabilizing methods are available
- What are the needs of search, rescue, and medical function

The S. Spec. is trained to use this information to:

- Perform structure triage (quick prioritization) if many buildings are involved
- Provide detailed hazard assessment of individual structures including drawings
- Work with search and rescue leadership to provide alternate strategies to reduce risk
- Design and help provide mitigation measures

It is relatively easy for engineers to recognize and report on the hazards, but it is much more difficult to devise creative, immediate ways to mitigate them. The goal is to have each task force staffed with S. Spec.'s, who will become trusted voices in determining the best course of action, and not just engineers that provide negative assessments. The level of acceptable risk may be high during the initial hours of a disaster but will change with time. Most hazard mitigation will slow rescue, thereby placing great pressure on the engineer/rescue relationship. Mutual trust and understanding are essential.

CHAPTER 6

LOCAL, STATE, AND FEDERAL RESPONSE

Oklahoma City Fire Department's (OCFD) Red Shift was on duty at 9 A.M. on April 19, and members of Station 1, five blocks from the Murrah building, were preparing to leave for a training exercise. They responded to the site within seconds, but because of the heavy smoke, they first reported the incident as a large fire in the parking lot just north of the Murrah building. As the fire was suppressed, the building damage became evident. Rescue Command plus two triage locations were set up, as OCFD, assisted by the Oklahoma City Police Department and local civilians, began rescue operations.

It must be stated that the initial, local response was extremely competent. Over 100 separate incidents were reported to the OCFD within minutes of the blast, including 50 vehicle fires. The OCPD responded immediately to assist with rescue operations, and establish perimeters and crime scene integrity. A total of 66 ambulances responded, including local, mutual aid, and self-dispatch and 204 patients were transported to the hospital within the first hour. All live and dead victims were removed from floors three through nine within the first 14 hours. As determined by subsequent rescue operations, the last live victim was located in the building at 7 P.M. and extricated by 8:30 P.M. (just prior to the initial deployment of the initial FEMA US&R task force).

The OCFD welcomed the support of the FEMA task forces, knowing that their specialized training and equipment would greatly aid the efforts in locating and rescuing the potentially large number of entombed victims. The federally assisted rescue effort that followed, although equally competent and well coordinated, did not achieve the positive results of live victim removal. It did proceed, under overall OCFD command, to locate and remove all victims from this dangerous environment without significant injury to rescue personnel.

By 9:10 A.M. State Department of Civil Emergency Management (ODCEM) staff began to arrive at the Emergency Operations Center (EOC), where they set up 24 hour operations. Those represented were ODCEM management; Oklahoma Departments of Education, Public Safety, and Health; the Oklahoma National Guard; Civil Air Patrol; and the National Weather Service.

Governor Frank Keating verbally declared a state of emergency shortly after 9:30 A.M. The FBI and ATF personnel were dispatched from Fort Worth to Oklahoma City. At 10:00 A.M., the Oklahoma National Guard was activated to provide security and the governor and his staff began operations at the EOC. Shortly thereafter, an incident command post, an FBI command post, and an on-site forward logistics and staging command post opened and began operations. The American Red Cross opened a shelter for the Regency Towers residents, and they subsequently provided food, water, and general assistance to disaster workers and victims.

The director of FEMA was charged with overseeing all non–law enforcement, and at 10:35 A.M., he activated two US&R task forces and a six-member incident support team (IST). The next day an additional four task forces were activated, and they were relieved by an additional five TFs during the second week. The IST, whose staff was quickly increased to more than 20, provided management of the TFs and coordination with OCFD command, who remained in control during the entire incident.

At 2:05 P.M., FEMA Region VI personnel arrived at the state EOC and began operations. The FEMA staff, both on-site and from headquarters, provided coordination and support to the state of Oklahoma and ODCEM.

At 4:00 P.M., President Clinton announced that he had signed Emergency Declaration FEMA 3115 EM. Plans to open a disaster field office (DFO) were initiated. Shortly thereafter, the first of the two US&R task forces arrived along with the first six members of the IST.

On April 20, the Myriad Convention Center was set up to accommodate the four additional FEMA US&R task forces that had, by then, been activated and were en route. The Oklahoma Restaurant Association, which was holding it's annual conference in the Myriad Convention Center, established a 24 hour food operation that would provide over 4,000 meals to rescue personnel. The community also generously supported the relief workers and families of the victims.

By 2 A.M., the FEMA Mobile Emergency Response Service (MERS) was supporting the FBI and other response agencies. Later on April 20, the Disaster Medical Assistance Team (DMAT) arrived from Tulsa, and the Public Health Services (PHS) mortuary team assisted and provided supplies to the medical examiner's (ME) office. The FBI special-agent-in-charge also established priorities as: (1) locate survivors, (2) remove victims, and (3) process the crime scene.

On April 21, at the request of Oklahoma City, the ODCEM and FEMA assisted the city in the opening of a multi-agency coordination center, (MACC) at the Myriad Convention Center, where a joint information center (JIC) had already been established. At that time, FEMA's disaster field office (DFO) was operating from a hotel until a site could be found.

On April 22, the DFO was operational in a new location that included the following emergency response teams (ERTs): communications, public works and engineering, firefighting, information and planning, mass care, resource support, health and medical services, and US&R. A congressional liaison office was also located at the DFO as well as the ESF5 (information and planning).

Later that day, the U.S. Small Business Administration (SBA) established a declaration for damage applications. The FEMA donations coordinator support team arrived and, in addition, the 54th Quartermaster, Graves Registration Unit, arrived to support the Oklahoma City ME's Office. At this time most response and recovery agencies were in place and operating.

CHAPTER 7

HAZARD MITIGATION ISSUES

As in the case of most building collapse disasters, the initial victim rescue efforts proceeded without too much regard for stopping to mitigate risk. There was a secondary bomb scare that temporarily halted the rescue efforts, but significant hazard mitigation did not start until the second day (April 20). Where bracing and shoring could be installed without slowing the rescue efforts, it was easily justified and proceeded without much discussion. Some projects, such as removal of hanging slabs, would cause curtailment of adjacent search and rescue, and, therefore, became the future source of conflict between engineers and rescue.

Early on day two, a six person Incident Support Team sent by FEMA to coordinate the federal US&R response met with the Oklahoma City Fire Department leadership to establish the initial strategy for what became a 16 day incident. The elements related to engineering were as follows:

- The IST would have an engineering component to aid, coordinate, and document the hazard mitigation efforts.
- The initial two task forces (Phoenix and Sacramento) would be reinforced with four others to have the three full TFs working in 12 hour shifts around the clock. (It was later decided that fresh TFs would relieve the initial six, and 11 total were eventually deployed.)
- Local volunteer contractors would continue to play a vital role in hazard mitigation and removal of debris.
- All significant decisions would be cleared through and coordinated with OCFD command.

During the entire incident, the following *collapse hazards* were discovered in the remaining structure:

- Column G12 was cracked at the third floor but was still supporting eight levels.
- Columns F20 and F22 were supporting seven levels, four through roof, with levels two and three having been stripped away.
- Columns F16 and F18 were poorly connected to the second floor beams, and they were also supporting eight levels above.
- The east end of the structure was disconnected from the stair/elevator bracing walls and was only marginally stable.

The following *falling hazards* were also present:

- A large, 16,000 kg (35,000 lb) section of roof beam/slab was hanging by a few rebars from the top of column E24.
- Numerous sections of concrete floor slabs and beams were hanging from the edges of the remaining structure.
- Many building contents were precariously perched near the edge of the remaining structure.
- Precast concrete panels that were suspended at the third floor over the south entry had their supporting beam connection partly broken and in danger of falling.
- The granite veneer panels on the east wall were badly broken, and some had fallen on the floor of the adjacent one-story concrete structure.
- Numerous unreinforced concrete interior wall partitions were left in leaning configurations by the blast.

As in any building collapse, the mitigation measures that were available were:

- Removal of the hazard
- Shoring or bracing—both vertical and lateral
- Monitoring hazard, accompanied by warning and safe haven/escape scheme
- Avoiding hazard

All of these methods were used in this incident. Most of the engineers who participated were contributors to the identification and/or mitigation of the many structural hazards.

The relative risk of further collapse and falling hazards was particularly difficult to assess in this incident. Task force engineers are trained with a heavy emphasis on evaluating earthquake damage, with the near certainty that aftershocks will continue to subject the damaged structure to lateral forces. Secondary collapse has been experienced by many structures after an initial seismic event, and lethal, falling debris is commonplace. This collapse is quite different from what most often occurs in earthquakes (see Comparison of Blast and Earthquake Effects on Reinforced Concrete Frame Structures, p. 26).

In the case of the Murrah building, significant secondary loading (in addition to gravity) was limited to wind pressures and forces generated by removal of large concrete debris. The structure had experienced severe damage and become dynamic, but had come to rest. The rebar that now served to suspend large pieces of concrete had probably received significantly higher stress in the process of stopping those sections from falling.

On the other hand, as the north half of the floor slabs separated at the building center (line F), the nominal rebar (temperature steel) in the north-south direction, which was placed (as is the normal case) just above the slab's principal bottom reinforcing, caused the slab bottom steel to be ripped out of the remaining slab for a distance of more than 1.5 m (5 ft) in some cases. This area of the remaining 152 mm (6 in.) slab, therefore, had reinforcing from its discontinuous top bars but no bottom reinforcing steel. In addition, the floor beams that remained were now spanning between the center and south wall of the building without the benefit of the balancing effect of their north spans. This could cause a significant increase in stress in the remaining beams' bottom reinforcing steel but could be monitored by observing crack patterns. These two conditions did not prove to be significant problems.

CHAPTER 8

INITIAL HAZARD ASSESSMENT AND MITIGATION

The Phoenix TF arrived on the evening of day one and performed the initial hazard assessment. During the second day, they directed the installation of pipe braces and confinement angles at the surviving, two-story column (G12) at the north wall. They started shoring operations in the first story, south of line F, in support of the rescue operations. They were then joined by the Sacra-

mento TF, who also began shoring and rescue in a nearby area of the first floor. A local volunteer contractor had previously installed some pipe braces at the third floor level of two columns (F20 and F22), and installed cables around damaged precast panels suspended above the south entry (see Figure 2.8).

The writer arrived in the early hours of the third day along with five engineers from the USACE. We became the IST engineering component and divided into two shifts of three engineers each. As dawn broke on day three (April 21), our task was to review the initial mitigation and become familiar with the building. After an extensive review of the building plans (several complete sets were made available), it was determined that much more work was needed in the following priority:

- Columns F20 and F22 needed more substantial bracing to avoid collapse.
- The east end of the structure was only marginally stable and needed to be monitored (no operations were ongoing within this area).
- Many falling hazards were hanging from the north edge of the remaining structure. The risk of their falling was less than in the case of a seismic disaster, but eventually this would need to be addressed.

During day three (April 21), the engineers from the four newly arrived task forces were briefed regarding the building's construction using existing plans on previously accomplished mitigation and the current mitigation effort. Arrangements were made to locate larger diameter pipes to more adequately brace columns F20 and F22, and a theodolite was borrowed from the Maryland TF-1 cache to monitor the east end of the structure.

On day four (April 22) it rained, which slowed operation. With the aid of LA Co-TF engineers and two firefighters, the IST engineering staff marked (numbered) all of the structure's remaining columns and prepared Mylar overlay drawings, color-coded to indicate the area of initial collapse, current hazards, and completed shoring. These drawings were available for future briefings and were updated as the incident progressed. During the morning briefing of the following day (April 22), the writer used the overlay drawings to more clearly define the important, remaining structural issues and suggest ways to mitigate them. This proved to be very timely, since during the previous night, an ill-advised attempt was made to cut down the large roof slab hanging from column E24.

After the April 22 briefing and subsequent daylong effort to deal with the large roof slab, an evening meeting of all S. Spec. and TF leaders, IST leadership, and lead engineers was demanded by the OCFD. The purpose of the meeting was to establish and prioritize an acceptable plan of action to deal with the hazards in concert with the very active victim recovery mode of the operation. The S. Spec.'s were instructed to coordinate their efforts, through twice daily meetings (at shift change), with the IST engineering staff. Inputs to modify the plan of action were to be coordinated through the IST lead engineer to enable the engineering staff to "speak with one voice."

Having a coordinated engineering opinion became even more important as the five additional task forces were rotated in, to replace the original six. The new S. Spec.'s, of course, had their own opinions, and as the incident moved into its second week, the risk/reward ratio shifted. Safety issues became more important as the hope for live victims faded. The new opinions needed to be filtered through engineering meetings and then discussed with operations leadership. The process worked well for the most part, but there were engineers who had trouble accepting the process when their own opinions did not prevail.

CHAPTER 9

DISCUSSION OF SPECIFIC HAZARDS

Columns F20 and F22

Because of the collapse of all floor levels to the north and floors two and three to the south, F20 and F22 (see Figures 2.3 and 2.4) columns were left carrying seven half-floors without any lateral support at the second and third floor levels. Instead of these concrete columns being braced and confined by the second and third floor beams, those floors had been ripped away, leaving cracked, uneven surfaces. The columns were somewhat restrained by the debris pile that surrounded them and extended about 7.6 m (25 ft) high on their north sides and less than 4.6 m (15 ft) on the south

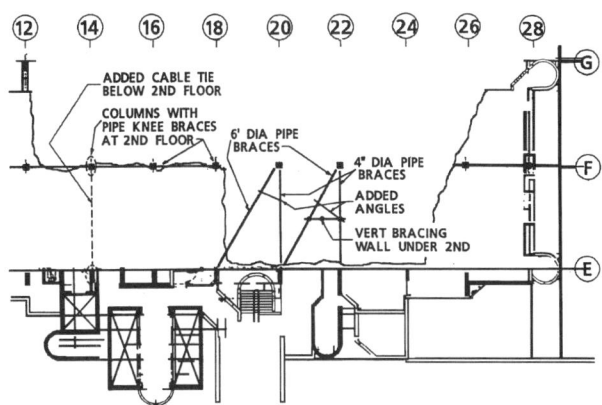

Fig. 2.3. Bracing of Columns

side (within the pit area). If the columns failed, however, they would do so suddenly, by buckling, and cause all seven currently supported slabs to be compressed together in a heap.

Initially, 4 in. diameter pipe braces were extended horizontally at the third floor level from the undamaged columns at line E and connected to columns F20 and F22. The 10.4 m (34 ft) long installed braces had a slenderness ratio (L/R) well beyond normally acceptable limits. In the writer's judgment, these columns were our highest priority collapse issue, and we concentrated our initial efforts on the design and construction of a more reliable bracing system. Some 6 in. diameter pipe in 12.2 m (40 ft) lengths was found that had a much greater compression capacity. Also, these columns needed to be braced in the east-west direction (as well as the north-south di-

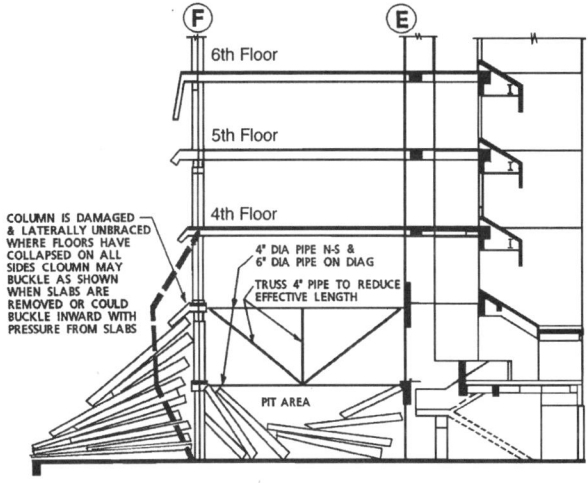

Fig. 2.4. Section at Column Lines 20 and 22

rection) at both the second and third floor levels. The pipes could be prefabricated on the adjacent entry plaza that led to the second floor entry and cause only minimal delays in rescue efforts in the pit during their erection. [Note that a 12.2m (40 ft) long, 6 in. pipe weighs 363 kg (800 lb), whereas the 4 in. pipe weighs only 204 kg (450 lb)]. The construction proceeded as follows:

- Four in. pipes were placed in the N-S direction at the second floor on lines 20 and 22, extending from E to F. This size pipe was used again, but they were trussed to the previously installed N-S pipes at the third floor to cut the effective length of both in half. Then the 4 in. pipes were also braced in the E-W direction.
- Since the collapsed slab rubble pile projected south of line F, it was not possible to place an E-W brace along line F. Therefore, 6 in. pipes were extended to both columns F20 and F22 at each floor, configured on a diagonal as illustrated to provide the E-W component of the needed bracing. Since the angle was 30°, the available E-W force was only 50% of the pipe capacity, but this was enough to provide sufficient bracing. (At a design level, an adequate column brace is normally considered as 1 or 2% of the column's vertical load.) However, the writer considers 2% as a bare minimum in the disaster environment. The diagonal configuration proved to be desirable, since it kept the braces away from the potential of being bumped during the removal of the collapsed slabs.
- The pipes were anchored to the columns at each end with four 19 mm (3/4 in.) drilled-in anchors. Unfortunately, some of the anchors did not hold during a subsequent slab removal operation. Then, 12.7 mm (0.5 in.) were installed with a loop around each column at each floor in order to guard against further pullout.
- To patch and better confine the concrete at each column where the floors had been torn away, an epoxy specialist was brought to the site who attempted to place epoxy mortar over wraps of small wire rope at column F20. Because of ongoing rescue operations and the inability to reach the north side of the column because of obstructions, this effort was only marginally effective. This contractor provided effective, epoxy in-

jection of the anchorage at line E to better guard against anchor pullout.

- On day 10 (April 28), a large slab adjacent to line 22 slipped onto the N-S second floor brace near column F22. It caused the brace to bend, and some of its anchors were partially pulled out. The brace was straightened with steel pipe shoring, and then a vertical pipe bracing wall was built under it and the adjacent diagonal pipe. Diagonal angles were added to the bracing wall to provide greater resistance from future shifting slabs.
- The final enhancements to this column bracing project occurred on day 11, when the continuing slab removal allowed access to all sides of each column at the third floor. The existing pipe brace connections were extended to form a full surround of each column, providing a more rigid tension connection than the cables, and 0.6 m (2 ft) long; steel sleeves were fabricated and placed over the cracked and poorly confined concrete at the third floor joints of columns F20 and F22. The sleeves were then filled with high strength, quick setting grout to form a structural bandage.
- These columns were monitored using smart levels (see Figure 2.9). The levels were placed on each of two surfaces between the second and third floors and set to zero. Any change in angle could then be measured with an accuracy of about one-tenth of a degree [lateral movement of 6.35 mm (1/4 in.) at each column joint]. The levels were read during the removal of all large adjacent slabs, using 10 × 50 binoculars from the safe area adjacent to line E, and no significant movement was detected for the remainder of the incident.

Columns F14, F16, and F18

The remaining second floor beams and slabs between lines F and E were cracked and nearly severed from the columns at line F (see Figure 2.5). Starting with the Phoenix and Sacramento TFs, this area was shored mostly using vertical wood-post shoring. The shoring was installed at the first story in this area (which became known as "the Forest") during the first days of the incident, using both 4 × 4 and 4 × 6 posts with strong x-bracing (see Figure 2.6). Several TFs installed this work, which was accom-

Fig. 1.7. North Side of Building after Bombing

Fig. 1.8. East Elevation

Fig. 1.9. West Elevation

Fig. 1.10. South Elevation (Building Front)

(a) (b)

Fig. 1.13 (a and b) Crane Lifting Beam from Debris

Fig. 1.14 (a). Removal of Column F24 from Debris

Fig. 1.14 (b). Close-up of Column F24 at Second Level Connection

Fig. 1.15. Columns F20 and F22

Fig. 1.16. Braces on Column F20

Fig. 1.17 (a). Third Floor Beam Spanning between F18 and E18

Fig. 1.17 (b). Cold Joint near Beam Midspan

Fig. 1.18. Column F16

Fig. 1.19. Remaining Portion of Transfer Girder

Fig. 1.20. Close-up of Transfer Girder

Fig. 1.21. Floor Slab Damage above Transfer Girder

Fig. 1.22. Underside of Typical Damaged Floor Slab

Fig. 1.23. Spalled Concrete in Roof Slab of East Annex Structure

Fig. 1.24. Damaged Precast Section on
Front Facade

Fig. 1.26. Athenian Restaurant Building

Legend

Structural Damage
Collapsed Structure
Broken Glass/Doors

Key

1 SW Bell
2 Journal Record Bldg.
3 Regency Towers
4 Post Office
5 Oklahoma Resource Board
6 Athenian
7 YMCA
8 St. Joseph School
9 A. P. Murrah Building
10 St. Joseph Cathedral
11 First Methodist Church
12 Federal Courthouse
13 Old Post Office
14 Oklahoma Gas and Electric Co.
15 Federal Reserve Bank

Data from: The City of Oklahoma City, Public Works Department, Map Prepared by
Geographic Information Systems, "Building Inspection Area," May 1995.

Fig. 1.25. Damaged Buildings in Vicinity of Explosion

Fig. 1.27. Oklahoma Water Resources Board Building

Fig. 1.28. The Journal Record Building

Fig. 1.31. The Regency Tower Apartment Complex

Fig. 2.6. Timber Scaffolding ("The Forest")

Fig. 2.7 (a). The "Mother Slab"

Fig. 2.7 (b). Cable Supports behind the "Mother Slab"

Fig. 2.8. Cable and Chain Supports for Precast Section

Fig. 2.10. "Diaper" on Hanging Slab

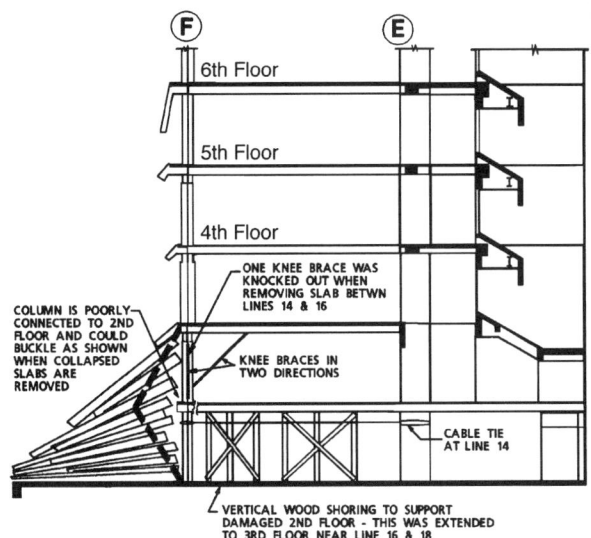

Fig. 2.5. Section at Column Lines 14, 16, and 18

plished not only to support the floor, but to allow the floor to provide lateral support for the columns on line F.

It became clear that the collapsed rubble slabs were leaning on these columns from the north, and that these slabs would eventually need to be moved. Therefore, more positive lateral bracing for the column needed to be designed. Knee braces, made from a 4 in. diameter pipe, were extended down from the third floor beams to provide a tension and/or compression force on two perpendicular faces of each column. Vertical shoring was also added between lines 16 and 18 to support part of the third floor, when it was observed that some of its cracks had increased in width.

On day seven (April 25), one of the knee braces along line F was knocked off by a large slab being removed by the crane. The Hilti anchors were pulled out of the column by the force, even though they appeared to have been installed properly. A new brace was installed using higher tension capacity anchors, supplied by the drilled-in anchors representative, and carefully tightened. (No torque wrench was immediately available, which is strongly recommended, as well as compressed air cans or other air jet devices to clean out the holes after drilling.) In addition to the enhanced anchorage, a cable stay was added to column F14 when the large slabs that were leaning against it were scheduled to be removed. Shortly thereaf-

ter, victim location data indicated that all potential victims had been removed from this area, and no additional nearby slabs needed to be moved. Therefore, no additional cable stays were installed.

East End of Structure

The east end of the structure had become structurally separated from the stair/bracing walls, and, therefore, was judged to be only marginally stable. Without the threat of aftershocks, the lateral forces that needed to be resisted by this section of the structure would be generated by wind. It had survived the blast with significant damage to the granite infill wall veneer (mounted on metal studs placed vertically between floor edge beams). The edge beams also were badly cracked at their connection to the northeast concrete vertical duct shaft. The shaft also had large duct openings at each floor and was badly cracked between the first and second floors. It was decided to monitor this section of the building by using a theodolite to sight from top to bottom of the northeast concrete duct shaft to determine if it remained vertical.

During the following three days, the maximum observed E-W movement at the top of this 37 m (120 ft) structure was only 12.7 mm (1/2 in.), which could probably be attributed to temperature change from one side of the cylindrical section to the other. Fortunately, the Oklahoma City winds remained relatively light (35 mph maximum) during the 16 day incident, and no larger than 19 mm (3/4 in.) movement was observed at any time. (Direct contact was maintained with the local weather service to warn of windy conditions.)

Column E24 was not laterally supported, except at the first floor and roof, because of the missing floors in that bay. It was monitored by another theodolite to observe any N-S movement. There were many pieces of beams with attached slabs hanging from this column, debris leaning to the south against it in the first two stories, and it was cracked at about the fifth floor. The theodolite revealed that the column bowed to the south, in the middle by about 12.7 mm (1/2 in.) [The line of sight remained within the outline of the 19 mm (3/4 in.) corner chamfer.] This condition remained constant throughout the incident. Since this section of the building was only accessible by crane or rope, it was initially searched and then avoided until the monitoring demonstrated that the probability of its collapse was relatively remote.

Large Section of Roof Beam/Slab (Slab from Hell or Mother Slab)

This 16,000 kg (35,000 lb) section of concrete was hanging by two 25.4 mm (1 in.) diameter, beam bottom bars from the top of column E24 (see Figure 2.7a). There were also several small slab bars still attached to adjacent slabs, but the primary forces were being carried by the two bottom bars. Since these bars were configured at approximately a 45° angle, forces of about 11,000 kg (25,000 lb) were acting on the column in the vertical and horizontal directions, pulling to the north. The horizontal force was being resisted by the remaining roof slab and transferred into the nearby stairwell bracing walls. The force also may have been helping to resist the collapsed concrete slabs that were leaning against the lower two stories of column E24. In any case, if the bars were cut abruptly, the 11,000 kg (25,000 lb) horizontal force would immediately change to zero with possibly catastrophic results to the adjacent, marginally stable east end of the building. During the night shift between days four and five (April 22–23), an attempt was made to remove this hazard by cutting the rebar and allowing the slab to fall to the rubble pile just north of column E24. The idea of removing this significant hazard was good, but this method could have caused a more serious problem than had been anticipated. Prior to cutting, the building had been evacuated, so a secondary collapse would not have injured rescue teams. However, if parts of the adjacent areas had collapsed, rescue efforts would have been seriously affected.

After several of the smaller bars and one of the two beam bottom bars were cut, the concrete slab shifted abruptly, so that the lower edge of the 508 mm (20 in.) thick beam section swung in to the west, and came to rest on the eighth floor near line 22 and about 3.05 m (10 ft) from line E. The cutting operation was discontinued at this point to reassess the situation.

At about 0800 hours on day five, the writers accompanied a group, which included OCFD, FEMA, Phoenix S. Spec. et al., to the ninth floor to assess the slab's current condition and devise methods for further mitigation. We found that the slab was still suspended from a single, 1 in. diameter rebar extending from one edge of the bottom of the concrete beam to the top of column E24 at the roof level. In addition, the lower end of the 16,000 kg (35,000 lb) slab was firmly bearing on the third floor. It was agreed that it was probable that the single rebar, because of its angle, was exerting a horizontal and vertical force on the column, and it would be unwise to proceed with cutting the bar. The possible results of cutting the bar could be the following:

- The 16,000 kg (35,000 lb) slab section could fall eight stories and severely damage anything in its path.
- The release of the potential 11,000 kg (25,000 lb) horizontal force at the column top would change the current force system in the roof and lead to unpredictable consequences, the most dire of which would be the collapse of the east end of the building.
- The collapse of a section of the eighth floor could occur when all of the weight of the 16,000 kg (35,000 lb) slab was released.

Since the goal was to remove this large falling hazard, we suggested that the slab be broken in place using normal cutting and small explosives with the pieces being caught in a dumpster suspended by a crane. OCFD summoned a trusted local explosive expert, who arrived at 4 P.M. the same day, and he advised that it would require from 24 to 30 hours to properly remove the slab using explosives. He further suggested that instead of using the explosive/removal scheme, the slab should be cable-tied to the stair wall on line 22. This could be accomplished much sooner. Because of the sensitivity regarding the use of explosives in this bomb-caused collapse and length of time that a major portion of the building would be restricted from rescue operations during the removal process, OCFD decided to allow the slab to be cable-tied and remain in place. The engineers were disappointed with the time estimate for explosive removal, but they could not argue with the OCFD decision based on the alternatives that were presented.

The slab was tied with two separate loops of cable during the night of day five. The loops passed around the slab and a section of stairwell wall between its north end and a window opening (see Figure 2.7b). There were several things about the cabling system that were less desirable than had been expected, such as:

- The cables were bent around 90 × 90 m wood strong backs on the slab instead of using steel pipe, which is not subject to cable cut-through and would have provided a smoother cable corner bend.
- The cables' configuration was essentially horizontal and did not provide any vertical lift to the slab. A more competent cable scheme was developed that included running additional cables through new holes drilled in the roof slab and then extended around the roof beam on line 22 near line E. This scheme was never implemented, since shortly thereafter the incident changed from rescue to recovery mode. (At that time, rescuers were excluded from working under this slab, and an excavator with a long reach was used to uncover the remaining victims in that area.)
- One of the cables was, in addition to being looped through the stair window opening, run over a precast wall spandrel panel at the third floor. The change in direction of the stressed cable caused a downward force in the panel and its supporting steel beam. Upon observing this at the beginning of the writer's shift on day six, the writers asked to have some solid wood shoring between the bottom of the north edge of the precast panel and the top of the concrete floor edge beam at line E.
- The cable clamps used to make the loops for attachment to the ratchet pullers were configured such that the middles of the three clamps had their U-bolts against the live side of the loop, instead of all three clamps being aligned with their grippers (saddles) on the live side. (Never saddle a dead horse.) The idea of reversing the middle clamp was explained as being better by the local contractor, since it gripped both sides of the loop, but I subsequently learned that it is not the safest way or recommended by the clamp manufacturer. (Another lesson was learned.)

The slab was monitored, using the theodolite, and personally checked by the writer and several others at least once a day for the remainder

Fig. 2.9. Smart Levels Used to Measure Angle Change in Columns

of the incident (see Figure 2.9). No significant movement was observed, but the slab did become a constant source of controversy regarding the risk/reward ratio of working under it. Regardless of how well an object as large as this is secured, it will prey on engineers' (and rescuers') minds until it is removed.

Numerous Sections of Hanging Floor Slabs

During windy spells that followed day five, small pieces of concrete and other debris fell from the slabs that hung from the north edges of the remaining floors. Much of the easily removable material was taken off and pulled back, beginning on day two. The roofing membrane, free hanging rebar, and building contents were all reached at their individual floor/roof levels and either removed or pulled back.

During day six, Los Angeles County TF leaders suggested that a careful inspection of the relative hazards involved with each of these concrete slabs should be made (which by now had been given names such as Australia, etc.). IST engineers agreed with their criteria for removal which were:

- Remove large (baseball size and larger) pieces
- Remove slabs where the bar that was hanging them could not be seen to pass through them
- Remove slabs that were suspended by bottom N-S rebar (temperature bars), since one could not be assured where they were spliced

After the completion of their survey of most of the hanging slabs on day seven, a short meeting was held in order to obtain IST operations and OCFD concurrence in removing the slabs. As a result, a group was assigned to review the survey and quickly came to agreement on which slabs required removal. The meeting occurred after our shift end, so the four of us remained through part of the night and were raised in a crane-lifted man bucket to physically mark each slab. Night rescue operations were then suspended on the north side of the building to concentrate the effort of both cranes on the rigging, cutting, and lifting of slabs, or cutting, and dropping them into a suspended dumpster. The work was nearly completed by sunrise.

This was a good example of how S. Spec.'s that were respected and well-integrated within their TF could work through the prescribed chain of command to resolve the conflicting issues that often accompany hazard mitigation. As a follow-up comment, it was ironic that after these slabs were removed, a piece of 6 in. thick insulating concrete from the roof fell near one firefighter during a subsequent wind squall. It weighed about 4.5 kg (10 lb), fell approximately 30 m (100 ft), and could have caused injury. The engineers had inspected it along with the other slabs. However, since the wire mesh reinforcing appeared to be restraining it, the engineers judged it to be a reasonable risk. Engineers are not infallible; they only do their best at the time. The remaining hanging slabs, those hanging below the "Mother Slab" at column E24, and those hanging from the east section of the building were later mitigated by the Menlo Park TF. They devised some unique cable stays and enclosures (diaper) of the larger slabs under the "Mother Slab" that could not be reached by the crane (see Figure 2.10).

CHAPTER 10

ENGINEERING SUMMARY

In total, 37 engineers participated in this incident (22 TF and 15 IST). This was a group effort; each worked as well as he could to help achieve positive results. From the initial assessment to the final hour, TF engineers supported their group's individual rescue task and then passed on their unique viewpoints to their replacements. The IST was staffed by dedicated engineers from the USACE, who willingly performed the monitoring and reporting functions. Additional USACE engineers were deployed to aid in the design of shoring in the nearby buildings where FBI safe access was required.

The entire wood shoring effort was monitored, mapped, and kept consistent by a New York firefighter/shoring instructor. The shoring method used had just been incorporated into rescue training. Taken as a whole, the emergency shoring at the Murrah building was superior to any previously constructed in US&R.

Engineers acted responsibly in attempting to minimize risk without hampering the rescue efforts. Most engineers became more uncomfortable with the uncertainties of some hazards as the days passed without the reward of finding live victims. All were better at identifying hazards than assessing their probability and devising creative, efficient solutions.

CHAPTER 11

BUILDING COLLAPSE EVALUATION GUIDE

Since disasters involving building collapse come in all sizes, from large earthquakes and hurricanes to individual building incidents, a responding engineer must have a clear picture of his purpose. Immediate analysis and advice is required from engineers that are trusted members of an emergency response unit by contrast with the more traditional engineer's role of observation, quiet reflection, and recording lessons learned. In either case, one must realize that although there may be some level of chaos at the scene, in most cases the local fire department will be in charge and operating under the incident command system (ICS). Task force engineers are trained to function within this system, but others will need to determine if and where a DFO is located, who is in charge, where to report, etc. Self-dispatched persons, however well intentioned, are often a liability to local authorities and are dealt with accordingly.

The collapse scene is most always loaded with images that will initially overload the engineering mind. The engineer that is well prepared for this experience will be most effective, and the ill prepared may suffer from the experience. Superior communications skills are essential, and experience, even with only table-top training problems, is extremely helpful. Assessment should be done by a minimum of two persons: a structural engineer familiar with structural collapse and a Hazmat specialist/first responder familiar with Hazmat identification and mitigation methods. Both individuals should not enter a confined space at the same time, and only one should enter without permission of the incident commander (IC). It is further suggested that no entry should be made without having a hand-held radio on command rescue frequency or personal locator device. In addition, everyone on the scene must know the warning signal system. (FEMA signals are: one blast = STOP work and listen for instructions; two blasts = O.K. to proceed; and three blasts = OUT—OUT—OUT! evacuate immediately.)

After the initial assessment, the real work of the "collapse engineer" begins. In cases in which trapped victims are involved, all the alternatives of hazard mitigation must be carefully explored: removal, brace/shore, avoid, and monitor (with appropriate evacuation plan/safe haven). The engineer must be prepared to design and carefully explain the least risky approaches so that the IC (and his associates) can assess the risk/reward ratio. Even the most difficult problems have some solution, and the engineer, in these pressure packed situations, must remain focused on his roll as the provider of essential information, and not the discussion maker.

The following is presented as an outline of procedures and equipment that, hopefully, will enhance one's effectiveness.

Collapse Structure Hazard Assessment Procedures

Necessary Equipment

- safety goggles with light and/or flashlight
- hard hat (be sure it fits well)
- dust mask, gloves, and knee pads
- comfortable boots with steel toes and shank
- rain gear and coveralls
- cash, personal identification, credit cards, etc.
- personal hygiene and medication items
- eyeglasses, safety glasses, plus extras
- clipboard, waterproof notebook, pens/pencils
- tape measure, compass, knife tool
- field glasses, geology hammer
- equipment belt or fanny pack
- personal field guide (ATC-20-1 plus helpful information for shoring, anchors, etc.)
- camera with extra battery
- electronic level, plastic strain gauge, marker to draw "X" through cracks
- small portable radio, tape recorder
- Hazmat detection and protection equipment (should be determined for each incident depending on types of Hazmat present)

Reconfirm cause of collapse—**Assume Nothing**

Identify structural type or types (use prepared hazard assessment form if possible)

Obtain plans or draw crude plan/cross sections

Identify basic vertical and lateral load resisting systems, brittle, ductile behavior and redundancy

Establish address, building orientation (side 1, 2, 3, 4, etc.), number of stories, basement, adjacent attached structures, establish column grid

Identify occupancy type and possible locations of Hazmat

Locate original ingress/egress: stairs, elevators, dumbwaiters, large ducts/shafts/mechanical interstitial spaces

Identify types of collapse, draw diagram, location of voids

Identify and grade potential collapse and falling hazards, relative risks of each based on aftershocks, wind, other possible forces, vibrations

Identify Hazmat issues and potential issues

Establish most probable victim location

Develop initial hazard mitigation strategy and discuss with appropriate leadership

CHAPTER 12

COMPARISON OF BLAST AND EARTHQUAKE EFFECTS ON REINFORCED CONCRETE FRAME STRUCTURES

Blast	Earthquake
Blast pressure damages all adjacent structure that is susceptible. Floor slabs and beams are usually extremely vulnerable to upward pressure and may be completely shattered. Weaker columns may be blown away, but larger, heavily loaded columns most often are not initially shattered.	Shaking damages dynamically brittle, vertical supporting elements (columns, short wall piers). Floor slabs and beams usually have little initial damage.
Blast pressures radiate from the point of detonation but decay very rapidly with distance and time. As the shock wave passes over a building, the pressures may change direction.	Shaking affects the entire structure and damage will occur because of mismatch in strength/stiffness ratio. Irregularities, both vertical and/or horizontal, will focus the damage to those most vulnerable areas (soft stories, short col.) Shaking may last for more than one minute.
Gravity acting on the damaged structure will cause it to seek a new state of stability.	Gravity acting on the damaged structure will cause it to seek a new state of stability.
The lack of lateral support, because of shattered floors can lead to buckling failure of adjacent columns and then to the collapse of one or more bays of the structure. If columns are shattered by blast, collapse of its supported floors will be likely.	The lack of vertical support, coupled with gravity acting on the heavy, relatively undamaged floor, will tend to cause a pancake type collapse of one or more stories.
Parts of the structure more distant from the blast may survive intact and have the potential of supporting one edge of some of the damaged floor slabs.	Earthquake damage can be focused in specific sections of the frame because of vertical and/or horizontal irregularities (soft story, short column, torsion, etc.). Partial collapse may then occur with slabs being draped from the remaining structure.
Blasts detonated in basements may cause even greater damage because of the initial confinement of blast pressures.	Earthquake shaking rarely affects basements since much smaller inertia forces are generated and basement structures are normally surrounded by strong, stiff, concrete walls.
Survival of victims is unlikely because of the initial effects of the blast and the initial shattering of concrete floors leads to relatively compacted, less survival voids.	Survival voids are often created in even complete pancaked collapse since the relatively strong floor slabs can bridge and drape over projections.
Secondary collapse is possible, especially if rescue operations require removal of collapsed slab structures that have become the temporary lateral bracing to remaining, free standing columns.	Aftershocks will cause additional lateral loading, which usually leads to some sort of readjustment to the structure (secondary collapse).
Victim removal may be accomplished by relatively easy removal of the shattered concrete, after appropriate stabilization of the remaining structure has been accomplished.	Victims are most often accessed by cutting thru the relatively solid floors (from the top) in a very labor-intensive process. Local and overall stabilization may be required.

PART III. DEFENSIVE DESIGN PRINCIPLES

CHAPTER 13

THREAT DEFINITION

The vehicle bomb is only one of many attack modes that may be directed against a facility or individuals. Some other forms of attack include: mob attack, kidnappings, poison gas attack, sniper attack, and mail bombs. However, from the standpoint of structural design, the large scale vehicle bomb governs design because, historically, it has caused the most structural damage and causalities. Also, designing to mitigate the effects of a vehicle bomb will automatically provide a certain level of protection against lesser threats, such as ballistic attack by rocket launchers, shape charges, etc. In addition to the vehicle bomb threat, another attack that is often of concern is a hand delivered small internal bomb within the lobby, loading dock, mail room, or other vulnerable area.

It could be argued that threats even more devastating than the vehicle bomb may be conceived, such as aerial or nuclear attack, but these threats are so structurally devastating to buildings that there is virtually nothing that can be done to mitigate the effects for civilian structures. On the other hand, there are a number of useful design options available for mitigating the effects of vehicle bomb attacks without making these structures into bunkers or fortresses. Other threats, such as chemical or biological warfare, are neglected in this discussion because they do not cause structural damage.

Although it is possible that the threat may change in the future, bombings have historically been a favorite tactic of terrorists for a variety of reasons, and they are likely to continue in the future. Ingredients for homemade bombs are easily obtained in the open market as are the techniques for making bombs. Also, bombings are easy and quick to execute. Vehicle bombs have the added advantage of being able to bring a large quantity of explosives to the doorstep of the target, undetected. Finally, the dramatic component of explosions, in terms of the sheer destruction they cause, creates a media sensation that is highly effective in transmitting the de-

sired message of the terrorist to the public (Kupperman and Trent 1979).

There are two parameters that need to be defined for a design threat: (1) the size of the weapon measured in pounds of TNT equivalent, W; and (2) its distance from critical structural components, or standoff, R. The size of the weapon used in design calculations is usually a decision made by the owner in collaboration with the security and blast consultant. The decision is usually based on a trade-off between the largest credible attack directed against the facility and the design constraints of the project. In general, the larger the threat considered, the more costly the protection and bunker-like the design.

A third useful parameter is the height of burst (HOB) of the weapon, which is the distance from the center of gravity of the weapon to the ground. For a vehicle bomb, in which the explosive is resting in the back of a truck, this is typically 0.91 m (3 ft).

For design purposes, large scale vehicle bombs typically contain hundreds to several thousand pounds of TNT equivalent, depending on the size and capacity of the vehicle used to deliver the weapon. Hand carried explosives are typically in the tens of pounds of TNT equivalent. The upperbound for a briefcase bomb is about 20 lb (9.0 kg), which is roughly the weight that a person can carry comfortably. For weapons delivered by hand trucks, the amount can be greater. Mail bombs are typically less than 10 lb of TNT equivalent. Unfortunately, there are few unclassified sources giving the sizes of weapons used in attacks. However, security consultants possess valuable information that may be used to evaluate the range of likely charge weights.

The critical weapon location to be considered in design is a function of the site, the building layout, and the security measures in place. For vehicle bombs, the critical locations are taken to be at the closest point that a vehicle can approach on each side, assuming that all security measures are in place. Typically, this is a vehicle parked along the curb directly outside the facility or at the sallyport where inspection takes place. For internal weapons, location is dictated by the areas of the building that are publicly ac-

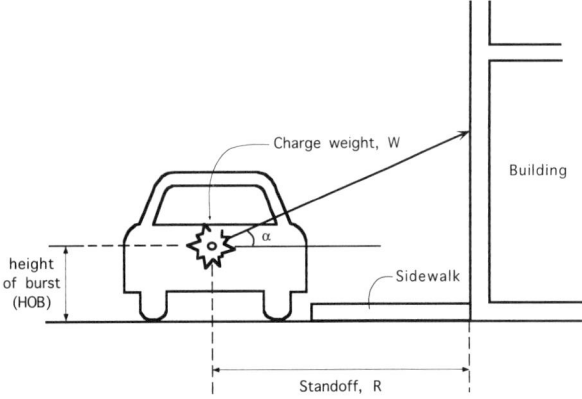

Fig. 3.1. Threat Definition

cessible, such as lobbies, corridors, loading docks, cafeterias, or retail spaces. Range or standoff is measured from the center of gravity of the charge located in the vehicle or other container to the building component under consideration (see Figure 3.1).

CHAPTER 14

WEAPONS EFFECTS

An explosion is a chemical reaction that causes an extremely rapid release of energy. This energy is released in several forms including sound (i.e., a loud bang), heat and light in the form of a "fire ball," and a shock wave that propagates radially outward from the explosive source at supersonic velocities. Of these effects, it is the shock wave, consisting of highly compressed particles of air, that is the primary cause of the structural damage.

Air-blast pressures are usually several orders of magnitude higher than ordinary loads for which the building is designed. Fortunately, these pressures only act for a fraction a second (or milliseconds) on the building. Because of the short duration of the loading, it is possible to design structures to withstand explosive forces.

In the wake of this compression wave or shock wave, there is a wind that is composed of air particles rushing in to fill the vacuum left by the shock wave. This is referred to as the dynamic or drag pressure, and it is responsible for carrying debris from the building site. It is be-

cause of this dynamic pressure that debris is found very far from the source of the explosion.

Part of the energy of the explosion is also imparted into the ground, causing a crater to form and a portion of the shock wave to propagate through the soil. This wave acts like a short-duration, high-intensity earthquake loading on the foundation of the building causing it to sway back and forth. This effect may or may not be of importance depending on the proximity and magnitude of the explosion.

Given the charge weight and standoff, the air-blast parameters may be determined using the charts found in military handbooks or software (see the bibliography section). These parameters are then used to develop loading functions for the various target surfaces to evaluate the appropriate design or evaluate the likely failure modes.

Explosive pressures decay very rapidly with time and distance. Temporally, pressures decay exponentially and spatially with the cube of the distance.

To put these quantities into perspective, a typical weapon often used to design military bunkers contains roughly 1,000 lb TNT equivalent. Given this perspective, it is not reasonable to expect an office building to survive as well as a military bunker, which is often underground, mounded with earth, and without windows. It is for this reason that our goals in providing protection are relatively modest, i.e., limited to preventing catastrophic collapse and to reducing the flying fragment hazard.

CHAPTER 15

DAMAGE MECHANISMS

The shock wave exerts extremely high pressures on all of the surfaces that it encounters in its path. The face of a building immediately in front of the face of an explosion is hit worst by the shock wave, not only because it is closest to the explosion, but because the wave is reflected from this surface, which amplifies the pressures by up to a factor of 13. This effect is analogous to a wave crashing against a retaining wall at the seashore. The wall stops the wave and forces it to reverse its direction. In this process, the wave jumps up on the wall. Similarly, the pressures jump very high on the side of the building

Sequence of Structural Damage

❶

Windows, exterior walls, columns and spandrel beams blown inward.

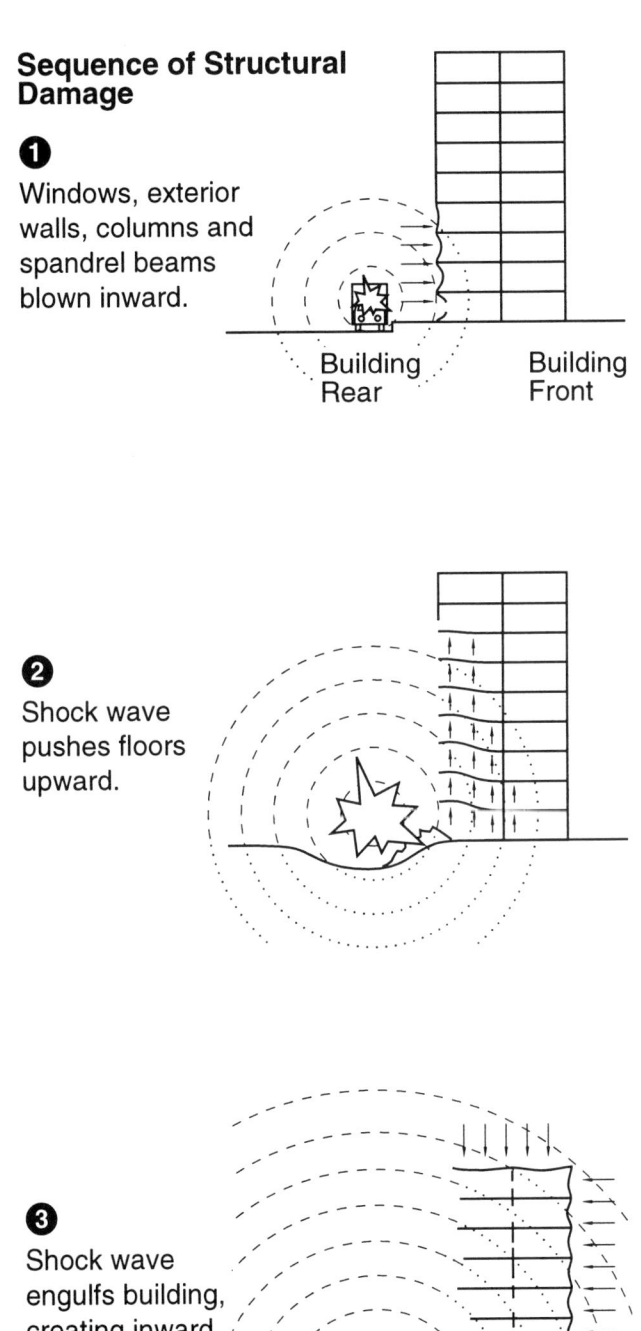

Building Rear   Building Front

❷

Shock wave pushes floors upward.

❸

Shock wave engulfs building, creating inward pressure on roof, sides and building front.

Fig. 3.2. Sequence of Structural Damage

that is impeding the path of the propagating wave, forcing it to reverse its direction or to be reflected.

The sequence of damage to a building subject to explosive forces is shown in Figure 3.2. The outer envelope of the building responds to the shock wave by deforming inward, causing windows to break and the exterior wall to fail (possibly). As the shock wave continues to expand, it enters the building acting both upward and downward on the floor slabs of the building. Since the floor slabs have not been designed for upward load, they are particularly vulnerable to collapsing in this loading direction. As the shock wave engulfs the building, the other sides of the building are subject to incident shock wave.

If the force of the explosion is capable of failing a floor slab or column on the lower floors, then a partial or complete collapse of the structure may be initiated as the floors above come tumbling down. Also, the collapse may propagate laterally as adjacent slabs and columns become destabilized by the removal of their neighbors, causing progressive collapse.

CHAPTER 16

DESIGN OBJECTIVE

The objective of providing protection to buildings vulnerable to terrorist attack is typically a function of the assets contained within the facility. For an office building, in which the primary assets are the occupants, the primary objective is to save lives. For power or telecommunications facilities, the primary objective is to maintain function. For facilities in which the contents are the chief assets, such as warehouses, the primary objective is to maintain the value of the contents.

In this discussion, the primary focus will be the protection of buildings in which the occupants are the primary assets. In this case, the objective of the structural engineer is to save lives by mitigating the structural damage of the building and reducing the chances of catastrophic collapse of the building. A secondary objective of protective design of these types of facilities is to maintain emergency functions until evacuation is complete. This may be accomplished by providing redundant emergency functions that are adequately separated. Also, placement of the emergency functions away from vulnerable areas, such as underground parking garages, is another consideration.

CHAPTER 17

SYSTEMS DESIGN

In discussing physical security measures implemented by architects and engineers, it is worthwhile to briefly put these countermeasures into the context of the overall security functions of the facility. Security measures can be divided into two groups: (1) physical and (2) operational. The distinction between these groups is that physical or passive security measures do not require human intervention (i.e., barriers, bollards, planters, structural hardening, etc.), whereas operational or active security measures do require human intervention (i.e., guards, sensors, closed circuit television, and other electronic devices). To have a balanced design, both these types of measures need to be implemented into the overall security of a facility. Any security system is only as strong as its weakest link. Architects and engineers can contribute to an effective physical security system that augments and facilitates the operational security functions. If security measures are left as an afterthought, expensive, unattractive, and makeshift security posts are the inevitable result.

If properly implemented, security countermeasures will contribute towards:

Preventing an attack. By making it more difficult to implement some of the more obvious attack scenarios (such as parking a truck along the curb adjacent to a building), the would-be attacker may become discouraged from targeting the facility. On the other hand, it may not be to our advantage to make the facility too obviously protected, because this may motivate the potential terrorist to escalate the threat to a higher level.

Delaying the attack. If an attack is initiated, through proper design, the architect and engineer can use devices to delay its execution by making it more difficult for the attacker to reach the intended target. This will give the security forces and authorities time to mobilize and, ideally, to stop the attack before it is executed. This is done by creating a buffer zone between the publicly accessible areas and the vital areas of the facility by means of an obstacle course, a meandering path, and/or a division of functions within the facility.

Mitigating the attack. If all these precautions are implemented and the attack still takes place, then the structure has been protected against such a threat. In addition, an emergency plan that considers evacuation routes, safe havens, rescue force requirements, and nearby hospital and triage facilities is provided. In the context of the overall security provided to the building, structural protection is a last resort that only becomes effective after all other efforts to stop the attack have failed.

An effective way to implement these goals is to create layers of security within the facility. The outermost layer is the perimeter of the facility. Interior to this line is the approach zone to the facility, then the building exterior, and finally the building interior. The interior of the building may be divided into successively more protected zones, beginning with publicly accessible areas, such as the lobby and retail space, to the more private areas of offices and, finally, the vital functions such as the control room and emergency functions. The advantage of this approach is that once a line of protection is breached, the facility has not been completely compromised. Also, by using this approach, not all of the focus is on the outer layer of protection, which may lead to an unattractive, fortress-like appearance.

CHAPTER 18

ARCHITECTURAL COUNTERMEASURES

There is much that can be done architecturally to mitigate the effects of a terrorist bombing on a facility. These measures often cost nothing or very little if implemented early in the design process. It is recommended that consultants are used as early as the site selection to optimize the protection provided. In the following paragraphs, suggestions are given for each of the lines of defense discussed in the preceding section.

Perimeter Line

The perimeter line of protection is the outermost line that can be protected by the security measures provided by the facility. In design, it is assumed that all large-scale explosive weapons (i.e., car bombs or truck bombs) will be outside of this line of defense. This line is defended by both physical and operational security methods.

It is recommended that the perimeter line be located as far as practical from the building exterior. Many times, vulnerable buildings are located in urban areas in which only the exterior wall of the building stands between the outside world and the facility. In this case, the options are obviously limited. Often, the perimeter line can be pushed out to the edge of the sidewalk by means of bollards, planters, and other obstacles. To push this line even further outward, restricting parking along the curb often can be arranged with local authorities. In some cases, street closings are an option.

Off-site parking is recommended for facilities that are vulnerable to terrorist attack. If on-site parking or underground parking is used, take precautions to limit access to these areas to only the building occupants and/or have all vehicles inspected. Place parking as far as practical from the building. If an underground area is used, consider placing the garage adjacent to the building rather than directly underneath the building. Another measure is to limit the size of vehicle that is able to enter the garage. This can be done physically by imposing a height limitation.

In some cases, barrier walls designed to resist the effects of an explosion can act to reduce the pressure levels acting on the exterior walls. However, they may not enhance security because they prohibit observation of activities that are occurring on the other side of the wall. In this case, an antiram knee wall with a fence may be an effective solution.

Access Control and Approach

Access control refers to the controlled access to the facility through the perimeter line. Architects and engineers can accommodate these security functions by providing adequate design for these activities, which makes it difficult for a vehicle to crash onto the site. This is done by means of barrier walls and other devices. Also, the location of access points should be oblique to oncoming streets so it will be difficult for a vehicle to gain enough velocity to break through these stations.

If space is available between the perimeter line and the building exterior, much can be done to delay an intruder. Examples include terraced landscaping, reflecting pools, staircases, circular driveways, planters, and any number of other obstacles that will make it difficult to rapidly reach the building.

Building Exterior Envelope

At the building exterior, the focus shifts from deterring and delaying the attack to mitigating the effects of an explosion. The exterior envelope of the building is the most vulnerable to an exterior explosive threat, because it is the part of the building closest to the weapon. It is also a critical line of defense for protecting the occupants of the building.

The design philosophy to be used here is: simpler is better. Generally, simple geometries with minimal ornamentation (which may become flying debris during an explosion) are recommended. If ornamentation is used, it is recommended that it consists of a lightweight material, such as timber or plastic, which is less likely to become a lethal projectile in the event of an explosion than, for instance, brick, stone, or metal.

The shape of the building can have a contributing effect on the overall damage to the structure. As an example of the effect shape can have on response, "U" or "L" shaped buildings tend to trap the wave, which may exacerbate the effect of the air-blast. For this reason, it is recommended that re-entrant corners are avoided.

The material used for the exterior wall needs to be carefully evaluated. Depending on the level of protection selected, the exterior wall can be designed to survive or fail in response to an explosion. If blast-resistant exterior walls are used, the loads transmitted to the interior frame should to checked.

On the other end of the spectrum are glass curtain walls, which are highly vulnerable to breakage and are not recommended. Unreinforced masonry, although it has substantial inertia, also has nominal resistance to lateral forces. If masonry is used, it should be reinforced both in the vertical and horizontal directions. Further discussion regarding exterior wall construction is given in Chapter 7.

Regardless of the construction material, the use of transfer girders is discouraged, because it limits the redundancy provided by the structural system. If transfer girders are used, then progressive collapse provisions that enable the girder to act as a catenary need to be incorporated.

Windows are typically the most vulnerable portion of any building. Although it may be impractical to design all of the windows to resist a large-scale explosive attack, it is desirable to limit the amount of glass breakage to reduce injuries. Annealed glass breaks at low pressure levels (0.50–1.0 psi), and the shards created by broken windows are responsible for many of the injuries incurred from large-scale explosive attack.

To limit this danger, there are several approaches that can be taken. One approach is to reduce the number and size of windows. If blast-resistant walls are used, then fewer and/or smaller windows will cause less air-blast to enter the facility, thus reducing the interior damage and injuries. Specific examples of how to incorporate these ideas into the design of a new building include: limiting the number of windows on the lower floors, where the pressures will be higher from an external explosive threat; using an internal atrium design with windows facing inward, not outward; and using clerestory windows, which are close to the ceiling, above the heads of the occupants.

In addition to studying the placement and size of windows, it is possible to use a more resistant type of glass or one that fails in a less lethal mode. Options in this category include the use of: tempered glass, laminated glass, or glass/polycarbonate security glazings. Tempered glass not only breaks at higher pressure levels, but it breaks into cube-shaped pieces that are considered to be less lethal. Laminated glass has been shown in recent tests to hold shards together and deform before breaking away from the frame. Glass/polycarbonate glazings are typically sold as bullet-resistant glazings and have the ability to resist high pressures. If polycarbonate is used on one of the exterior faces, then care must be taken to ensure that the surface is not prone to scratching, clouding, or discoloring. Each of these products has its advantages, and the manufacturers should be consulted prior to making a decision.

Windows, once the sole responsibility of the architect, become a structural issue once explosive effects are taken into consideration. When considering the installation of special window lights, the structural design of the mullions, frame, and the supporting wall need to be checked to ensure that it is capable of holding the window in place during an explosion. There is no point in designing a window that is more resistant than the wall that is holding it in place.

Two measures that may be considered for the retrofit of existing windows include: (1) polyester film coating on the inside face of a window and (2) blast curtains to catch the shards if the window breaks. Again, the manufacturer needs to be consulted regarding the tendency of this material to scratch, delaminate, cloud, or discolor. Blast curtains are a British invention that consist of Kevlar curtains that hold the shards but permit the air-blast to pass through.

Building Interior

The protection of the building interior can be divided into two categories: (1) functional layout and (2) structural layout. In terms of functional layout, public areas such as the lobby, loading dock, and retail area need to be separated from the more secured areas of the facility. This can be done by creating internal "hard lines" or by creating buffer zones, using stairwells, elevator shafts, corridors, and storage areas between public and secured areas.

Emergency functions and elevator shafts should be placed away from internal parking areas and loading docks, as we learned from the World Trade Center bombing incident. Elevator shafts can become chimneys in the event of an explosion, transmitting smoke and heat from the explosion to all levels of the building. This may hinder evacuation and increases the risk of injury from smoke inhalation. Emergency functions, such as sprinkler systems and generators, are critical for mitigating the effects of an explosion and they need to be placed away from vulnerable areas of the building such as underground parking areas.

False ceilings, venetian blinds, ductwork, air conditioners, and other equipment may become flying debris in the event of an explosion. Wherever possible, it is recommended that the design should be simplified to limit these hazards. Placing heavy equipment, such as air conditioners, near the floor rather than the ceiling is one idea for limiting this hazard. Using curtains, rather than venetian blinds, and using exposed ductwork as an architectural device are other ideas.

In terms of structural layout, it is recommended that the bays of the building are kept to dimensions less than or equal to 9.2 m (30 ft) within floor-to-floor heights less than or equal to 525 m (16 ft) as a method of good engineering practice. Since the outer bay is more vulnerable to collapse than the inner bays, it is worthwhile to consider reducing the depth of the outer bay and to use multiple interior bays to limit the collapse hazard. Proper reinforcement for these members may be determined by using dynamic structural analysis techniques.

Progressive collapse measures also need to be implemented to ensure that the damage is limited to the immediate vicinity closest to the explosive source. Some methods for limiting the progressive collapse in the design of new buildings are given in the references listed in the Bibliography.

Special attention should be paid to those areas that are most accessible to the public. Typically, these include:

- underground parking garages
- loading dock entrances
- lobbies
- mail rooms
- retail spaces

Each of these is discussed next.

In underground parking areas it is recommended that all nontenant vehicles be inspected prior to entering. Presuming that this is enforced, the weapon that is able to gain access to this area is expected to be considerably smaller than any threat exterior to the building. In these areas, it is recommended that the columns be designed to span at least two floor heights without buckling. This is recommended because often the floor slabs will be lost, resulting from weapons effects, leaving the columns unbraced for two or more levels. Emergency functions and elevator shafts should be placed away from internal parking areas and loading docks, as we learned from the World Trade Center bombing incident. Finally, it is recommended that office space in the basements be relocated to less vulnerable areas. One way to limit this hazard is to not use stairwells or elevator shafts that are continuous from the basement to the roof. Another way to reduce the vulnerability of an underground parking garage is to place it adjacent to the building, instead of directly above it.

Loading dock entrances increase the vulnerability of the structure immediately adjacent to them. In these areas, trucks are able to gain closer access to the building than along the secured perimeter. Also, in these areas, transfer girders are sometimes used to span the entrance. It is recommended that high population areas in spaces facing the entrance be relocated to less vulnerable locations, and/or the structural components be designed to withstand the direct effects of an air-blast in these areas. The use of transfer girders is discouraged. Also, hardened design of the roof and walls of the loading dock entranceway is advisable.

Lobbies are difficult to design to resist the effects of an air-blast. Typically, the threat of concern in this area is a briefcase bomb outside the inspection point. Often lobbies contain a lot of glass, which is hazardous and in open areas difficult to control. Hazard reducing strategies include reducing access to columns through architectural measures and using special hazard mitigating glazing, such as laminated glass.

Mail rooms can be a concern in some facilities. One option to circumvent this vulnerability is to presort mail at another site. If this is not possible, then partially contain the weapons effects by using reinforced (in both the horizontal and vertical directions), concrete block walls. To be effective, they should be properly anchored into the floors above and below. They should be used on three sides of the mail room. The fourth wall should be designed to vent the explosion towards the building exterior. Also, it is a good idea to remove populated areas immediately adjacent to the mail room to provide a buffer zone.

For retail spaces, or other public areas on the ground floor, it is possible to protect the occupants by designing a hardened envelope above and behind these spaces. As an additional precaution, relocate office spaces immediately above or below these areas. Also, special attention should be paid to supporting columns in these areas to prevent the initiation of progressive collapse.

CHAPTER 19

STRUCTURAL COUNTERMEASURES

Construction Materials

Generally, well designed cast-in-place reinforced concrete structures provide a significant level of protection against explosive loads compared with other construction materials. This is because of their heavy weight, monolithic character, and ductile response. Steel frame construction may also be effective in resisting explosive effects. For steel structures, the frame has considerable ductility if the connections are properly designed and constructed. However, the materials used for the floor system, the exterior wall cladding, and their connections to the steel frame often have less resistance to explosive effects. Consequently, the frame may survive but the occupants and contents may not. Reinforced masonry and precast concrete can also be designed for low levels of blast protection. Unreinforced masonry and prestressed concrete construction are discouraged.

Cast-in-Place Reinforced Concrete Construction

For reinforced concrete design to behave well under explosive loading, it is imperative that ductile connections be used to ensure that failure does not occur at these critical locations. For cast-in-place reinforced concrete design, the following minimum properties are recommended:

- 28 day compressive strength of concrete: 27.5 MPa (4,000 psi)
- yield strength of reinforcing steel: 413.7 MPa (60,000 psi)

Minimum dimensions recommended for cast-in-place reinforced concrete components are:

- floor slab thickness: 20.3 cm (8 in.) minimum
- rectangular column width: 30.5 cm (12 in.) minimum
- joist width: 10.2 cm (4 in.) minimum
- exterior wall thickness: 25.4 cm (10 in.) minimum

To ensure a ductile response and prevent progressive collapse, the following measures are recommended:

- two-way slab system supported by beams on four sides to provide an alternate load path
- ties placed along the entire length of beams and girders and columns to provide confinement
- continuous top reinforcement in slabs to resist the upward load
- continuous vertical reinforcement on both sides of exterior walls to increase the ultimate capacity of the envelope

- seismic detailing at connections ("Details" 1986)
- continuous reinforcement in floor systems or staggered lap splices that develop the full strength of the reinforcement
- proper anchorage of reinforcement bars
- continuous bottom reinforcement in slabs along column lines to prevent progressive collapse (Mitchell and Cook 1984)
- exterior and interior columns in public areas for unbraced lengths of at least two stories
- outer designed bay to resist progressive collapse initiated by the loss of a ground floor column or other primary support

One-way joint systems, waffle slabs, and flat slabs with drop panels can also be designed for modest levels of blast protection. Again, proper anchorage and shear reinforcement are required to make these systems effective.

Analysis Methods

Design of blast-resistant structural members is generally accomplished by performing non-linear dynamic analysis. Using static elastic methods usually gives overly conservative solutions, because they neglect the short duration of the loads and the ductile capacity of the members.

Dynamic analysis may be performed using single-degree-of-freedom, multi-degree-of-freedom, or finite element methods. The most common approach is to use single-degree-of-freedom methods. This is because these methods are provided in the military handbooks for analysis, and also this method is generally the most cost-effective in design situations. However, except for the simplest structural components, usually an experienced analyst is needed to perform these computations, because a high-level degree of engineering judgment is required in developing meaningful models and interpreting the solutions. A graphic summary of the dynamic analysis methodology is given in Figure 3.3.

The references given in the bibliography give detailed methods on how to design members for blast resistance. To give a flavor of the methodology used, a brief description of a simple design problem will be detailed. Consider the design of an exterior wall subject to an explosion. These loads will vary spatially

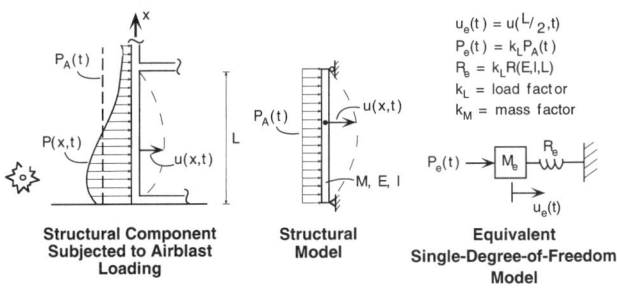

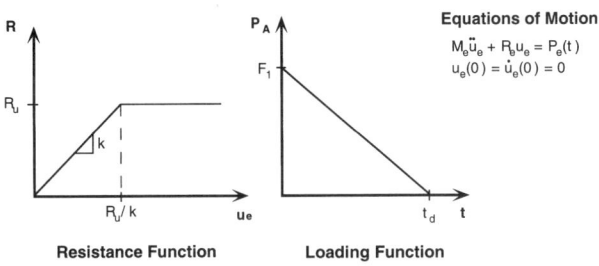

Fig. 3.3 Dynamic Analysis of Structural Components

along the member. For a single-degree-of-freedom analysis, usually this loading will be approximated by a spatially uniform loading by using the peak load or an averaged loading acting along the length of the wall. The time decay of the loading is usually taken to be a linearly, decaying function. Reduction is typically selected to conserve the impulse (i.e., the area under the pressure-time curve). The member itself will be modeled as a lumped mass and nonlinear spring. Usually the spring has the properties of an elastic–perfectly plastic material with the stiffness and ultimate capacity defined by the end conditions of the system. Equivalency factors for the mass, spring, and load are determined to give the same displacement as the actual system. Solution of this system is determined by using charts or numerical integration techniques. Numerous elaborations or variations are possible for this simple system depending on the particular component and loading situation under consideration. Results are evaluated usually by using a ductility or support rotation criteria. For finite element computations, a strain criteria may be used.

This example is appropriate for members that respond flexurally to explosive effects. For members that are close-in (i.e., $W/R^3 < 1$) or for

members with small length-to-depth ratios, shear is likely to be a governing mode of failure and needs to be checked.

CHAPTER 20

PROTECTING EXISTING BUILDINGS

Every building has some inherent ability to resist explosive effects. Depending on the construction materials and the structural design approach used, this resistance may vary significantly. A masonry or brick load-bearing building, for instance, has very little inherent blast resistance, whereas a poured-in-place reinforced concrete or steel frame building will have much more inherent strength.

Once the inherent strength of the building has been approximated, the effectiveness of various solutions can be evaluated.

Probably the best use of funds allocated to upgrading the protection of a facility is to consider ways of increasing the standoff to an external vehicle bomb by using perimeter security measures. Given the rapid decay of air-blast pressure with distance, this may have a dramatic effect on increasing the inherent protection provided to the facility.

Another popular method of upgrading the protection provided by a facility is to mitigate the glass hazard by providing polyester film coating on the windows; using blast curtains; or replacing the windows with security glazings, laminated glass, or tempered glass. This can be a costly option and the various solutions need to be weighed carefully in terms of the protection afforded. Also, remember that it is of little use to increase the resistance of the glazing above that of the wall system.

Structural options for upgrading the security at an existing building include: building a hardened wall behind the existing wall and replacing the existing exterior wall that is properly anchored into the existing structure. If this is done, it should be verified that the structural frame can adequately transmit the reaction loads from the exterior walls.

Often, for buildings with little inherent resistance, a structural upgrade is not an effective option. In this case, rearranging the internal functions of the building may be an option.

CHAPTER 21

ASCE ACTIVITIES

ASCE has been active during the past several years in educating the engineering community regarding the effects of explosions on structures. At the time of the writing of this book, two other publications are simultaneously being prepared by ASCE. One is an ASCE State of the Practice Report entitled "Structural Design for Physical Security," prepared by the ASCE Committee on Shock and Vibration. The other is by a group of ASCE engineers who traveled to Oklahoma City after the incident, who are preparing a report on their findings (*The Oklahoma City Bombing: Improving Building Performance Through Multi-Hazard Mitigation*, FEMA 277, Federal Emergency Management Agency Mitigation Directorate/ASCE, Washington, D.C., August 1996).

APPENDIX. OKLAHOMA CITY INCIDENT SUPPORT TEAM, STRUCTURES SPECIALIST LOG

19 APR 95	2000HR	AZ TF-1 arrives
	2300	CA TF-7 (Sacramento) arrives
20 APR 95	0600HR	CA TF-7 begin wood shoring of 1st story. Cols. F–E and 18–16 AZ-TF-1 begins work on north debris pile
	P.M.	Contractors help TF install vertical pipe shores at Col. G-12. Also install 4 in. diam. horiz. pipe braces at 3rd floor. Cols. E20–F20 & E22–F22
	2200HR	I.S.T. lead engineer + USACE struct. spec. arrive
	2400	Four additional TF arrive. CA TF-2 (LA Co), MD-1, NY-1, VA-2 (Va Beach)
21 APR 95	0600	Shift change Mtg—situation update (same for every day following)
	0700	Begin monitor East Tower col. lines 26–28. Separated sect. of bldg. Review plans of building
	0900	Meet with Boldt Const. Co. Discuss need for additional bracing at Cols. F20 and F22. Need braces at 2nd & 3rd floor in E-W & N-S direction + 4 in. diam pipes are too slender to be reliable compression brace. Boldt to try locate 6 in. diam. pipes or 6 in.^{2} tubes
	1100	Meet Ed Kirkpatrick (original engineer of building) discussed that bldg. relies on shear walls at stairs & elevator for lateral bracing. No frame action. He offered help in understanding design & left his beeper no.
	P.M.	Continue monitor East Tower and falling hazards Watch OPS in Pit area (Cols. E20, E22, F20, F22). Inform OPS to coordinate all removal of large concrete pieces from around cols. F20 & F22 Shift change meeting—Twice every day typical
22 APR 95	0700	I.S.T. engineers tasked w/preparing new hazard/shoring/bracing drawings. Plus to mark all building cols. w/grid lines & story
	0800	Contacted Boldt to obtain Mylar overlay blank sheets for drawings. Set up plan table area on loading dock (under leaky roof drain) Rainy & cold day—OPS slowed due to slippery conditions on debris piles— safety first—Gave opinion that rain would not add significant load to this structure—only concern was lubrication of slabs, rescuers and/or slabs might slip
	0900	L.A. Co. engineers help by marking all columns at each floor level. COE engineers begin to map collapse area of each floor
	1000	Receive Mylar and begin making overlay drawings w/help from L.A. Co.
	1100	Discovered that first floor is framed w/beams & slabs per typical floors. What is extent of basement—are we shoring over struct. slab or is 1st floor on grade. Tried to get into basement but found no access. Called E. Kirkpatrick—he clarified that floor is designed as framed Fl due to clay soil, but floor was poured over cardboard forms. Informed OPS that our shoring that rests on this slab is O.K.—nil risk of collapsing 1st Fl—Watch tightness of shims & report
	1200	FBI requests IST engineer to visit their bldg. at 11 N.E. 6th to evaluate its safety—Looked O.K. Only 6 windows broken—roof & 2nd floor O.K.
	1300	Continue draw overlays + 8 1/2 × 11 collapse zone/hazard zone of ea. floor
	1400	Evacuate due to bomb threat
	1600	Return & inspect shoring + record on overlays
	1700	Boldt locates & delivers 6 in. diam. pipes for Col. braces—discuss and agree on diagonal configuration to get N-S as well as E-W component of bracing
	2000	Finish hazard mitigation strategy ready for presentation in A.M.
	late P.M.	Continue support of OPS, Monitor East Tower & falling hazards

23 APR 95	0600	Bldg. evacuated in order for TF to cut & drop large (3,500 lb slab) from roof/9th floor (Mother Slab). Poor idea but crane cannot reach/lift slab and it is large hazard. OCFD I.C. allows TF to try it. Slab bars are partly cut and slab swings to rest its lower end to bear on 8th floor, line 22—operations stopped.
	0730	IST lead eng. presents hazard mitigation priorities & strategy using drawings
	0800	IST lead engineers, IST OPS, & OCFD I.C. meet at 9th floor, line 22 to assess Mother Slab—continuing problem from drop & cut attempt. IST engineers suggest slow controlled removal of Mother Slab, use explosives
	1000	Continue monitor East Tower w/theodolite + add col. marks to be better seen by crane operator & front OPS
	1400	Names of local explosives experts are obtained from local COE office. Gave 3 names to OCFD OPS. Dykron, Tulsa was called & arrived at 1700.
	1500	OCFD I.C. cleared Boldt to install additional pipe braces—col. F20 & F22
	1700	Dykron suggested he would need 24–30 hr to explosion demo slab—faster to tie it back to adjacent stairwell at E22 + remove all loose hanging pieces. OCFD says O.K. to tieback of Mother Slab & assigns job to Flintco for P.M.
	2200	Meeting of TF leaders/engineers, IST OPS & engrs, OCFD OPS called by OCFD Ch. Maars to define/clarify hazard mitigation priority/strategy. Establish hazards mitigation plan and require that engineering coordination meetings be held at each shift change (every 12 hr)—engrs. speak w/one voice
24 APR 95	0800	Inspect tieback of Mother Slab that was done during nite. Need to add cribbing under P.C. window sill panel where cable bends. Ask Flintco to remove more small pieces. MD TF-1 remove loose pieces on 8th & 9th floors
	0900	COE engr. go to brick buildings north of site for FBI—need to shore
	1000	Gave opinion that "No Fly Zone" could be lowered to allow normal landing pattern to Airfield—No full power takeoff over building needed
		Continue monitor East Tower w/theodolite
	1400	Advise rear OPS re. safety of 3rd floor slab lines E–F and 12–18
	1600	Work w/Boldt to get last 3 pipe braces in—need 50T crane for last 6 in. pipe—Coordinate w/IST OPS & OCFD
	1700	Calculate remaining debris gross volume on north face = 3300 c.y. (@25 c.y. each = 125 dumpsters)
	1800	Give suggested size of maximum lifts for slabs/columns. 100 s.f. slab = 7,500 lb; 15 ft beam = 1,500 lb; 1 story col. = 5,000 lb; 10 ft transfer beam = 22,000 lb
	1900	Met w/L.A. Co engineers + OCFD rep. to discuss which hanging slabs on north face needed to be removed. Agreed on most & got O.K. from OCFD & OPS
	2000	Marked all hanging slabs to be removed from crane basket—OPS to have them all removed by dawn
	2030	Got epoxy contr. started on doing wrap of col. F20 where 3rd & 2nd floors were ripped away. Will have to work around rear OPS
	2200	Discussed need for additional shoring of 3rd floor west of pit—TF S. Spec.
	2300	Discussed conveyor being placed on north face—faster debris removal
	2400	Provide Allied Crane OP Xerox copy of bldg. dimensions—size boom
	late P.M.	Discussed large vee slab w/TF Struct Spec—need to cut & split it to remove
25 APR 95	0700	IST OPS briefing to TF followed by S. Spec. briefing (typical each day on)
	0800	Continue monitor East Tower + hourly check of Mother Slab cables
	1000	Received 2 complete sets of building plans from Flintco. & COE
		Agreed on method to add lateral pipe bracing in N-S dir. col F22–E22 2nd
	1100	Discussed lifts of large conc. floor slab/beam sections N of line F
	1400	Discussed need for pipe bracing at cols. F16 & F18 identified by CA TF-2 S.S. Notified Boldt to move to F16 & 18 after F20 & F22 pipes are placed
	1500	Briefed on coming WA TF-1 S. Spec. about structure & started them on additional shoring of slab F16, F18–E18, E16—1st & 2nd story

	1800	All pipe braces complete in pit area
	1800	Typ. coordination meeting—no major issues
	late P.M.	Continue monitor and aid w/lifting large slabs
26 APR 95	0700	Typ. coordination meeting—no major issues
	0800	Epoxy bandage on col. F22 started but will conflict w/OPS—discontinue Discussed need to add tension cables to pipe braces F20 and F22 since drilled in anchor pulled out when lifted slab hit col.
	1000	Started cabling cols. F20 & F22. Discussed lifting of slabs on N. face w/Ops. Checked Mother Slab cables hourly
	1100	Advise recutting or tieback of slab at F22, 5th floor. Advice = remove
	1200	Epoxy bandages on col. F20 done at 3rd floor & part done at 2nd floor—discontinue
	1400	40 mph winds predicted. Increase monitor operation of East Tower. Max horiz. movement at + 120 ft was 0.62 in. quite stable
	1500	COE engineers to give aid to FBI in assessment of Okla. Water Bldg.
	1530	Receive two 50X spotting scopes from COE, Tulsa
	1600	Receive two theodolites from local civil eng. to replace wind damaged one
	1800	Typ. coordination meeting—no major issues
	late P.M.	Continue monitor and aid w/lifting large slabs
27 APR 95	0700	Typ. coordination meeting—no major issues
	0800	Continue monitoring of East Tower + also watch col. F22. ALL STABLE Checked Mother Slab cables hourly
	0900	Checked on progress of cable ties cols. F20 & 22. 20 was O.K. but hard to tighten since used turnbuckle—added WD40. F22 was not tight since they used choker at each end—N.G.—need to be changed to hard loops w/T. buckle
	1000	Advise on pipe braces being added at cols. F14, 16, & 18
	1200	Brief on coming S. Spec. from VA TF-1 & CA TF-3 (Menlo Park) Need to assign COE engineer to fill in for VA TF-1—COE will send additional S. Spec. to help other govt. agencies and rotate out original COE S.S.
	1300	Set Smart Levels on cols. F20 & F22.
	1400	Continue monitor & advice Work w/CA TF2 S. Spec. & Boldt to add cable tieback of cols. F14–E14 in 1st story—Beam at 2nd floor is shored & poorly connected to col. F14
	1500	Received binoculars to check Smart Levels from SAFE area at Line E (50X scope N.G. to view from 40 ft)
	1800	Typ. Coordination meeting—no major issues
	late P.M.	Continue monitor and aid w/lifting large slabs
28 APR 95	0700	Typ. coordination meeting—no major issues
	0800	Continue monitoring of East Tower + also watch col. F22. ALL STABLE Checked Mother Slab cables hourly
	0900	Worked w/Boldt to straighten & strengthen the horiz. pipe braces to col. F22 where cono. debris shift had leaned in. Designed & communicated to add 3 vert. pipe supports w/diag. brace in their plane from 1st floor slab to 2nd level
	1000	Set Smart Level at col. F18, 2nd story, movement was suspected
	1200	Swapped Smart Level Control units w/manufacturer's rep. in OKC (battery saver circuit turned off)
	Hourly	Checked Smart Levels on half hour intervals and during lifts of slabs off N. debris pile to held other govt. agencies and rotate our original COE S.S.
	1300	Gave tour to engineers hired by GSA to study future use of building/site
	1400	Continue monitor & advice
	1800	Typ. coordination meeting—no major issues
	late P.M.	Continue monitor and aid w/lifting large slabs

29 APR 95	0700	Typ. coordination meeting—no major issues
	0800	Continue monitoring of East Tower + also watch col. F22. ALL STABLE Checked Mother Slab cables hourly
	0800	Look at conditions near 2nd & 3rd floor joints at cols. F20 & F22 to design new surrounding collars so cables will not be needed (Cables are difficult to keep tight and may be too limber to resist col. buckling.)
	1000	Col. F22 at removed 3rd floor joint was inspected up close & cracks can be seen that may extend thru column concrete section. Try to find 1 or 2 in. banding
	1100	Add pipe shore to 4 in. pipe brace at col. F20 for potential of falling slab
	1130	Discuss & sketch collar joints at each end of pipe braces for cols. F20 & F22 w/ Boldt (Lex Paine). *He will get job done.*
	1200	Locate coaxial epoxy cartridges at base but caulking gun is missing
	1300	Found epoxy gun—Boldt starting on collars
	1400	Give advice on lifts + breakup of leaning slabs
	1500	Get permission of OCFD & call Flintco to make steel sleeves for cols. F20 & F22 at 3rd floor—to do 2nd floor when uncovered. Order nonshrink, fast set grout to fill collars & bandage col. joints. Grout = ROCA—15 min set & 4,500 psi in 2 hr
	1600	Monitor cols. F20 & 22 at each lift of large slab on north face
	1700	Flintco on job w/steel pls. & angles to start steel sleeves for cols. F20, 22 @ 3rd
	Hourly	Checked Smart Levels on half hour intervals and during lifts of slabs off N. debris pile
	1800	Typ. coordination meeting—Discussed collar & sleeve installation—follow
	late P.M.	Continue monitor and aid w/lifting large slabs Organize bucket brigade to pour grout in sleeves at F20 & 22 @ 3rd
30 APR 95	0700	Typ. coordination meeting—report grouting successful—collars ongoing Struct. spec. want new meeting to discuss Mother Slab
	0800	Continue monitoring East Tower + also watch col. F22. ALL STABLE Checked Mother Slab cables hourly + check Smart Levels when lifts occur
	0800	S. Spec. meeting on Mother Slab. Remove vs better tieback. R. Barrett, CA-TF3 (Menlo Pork) wants to review panel in field to design better tieback system
	0900	CA-TF3 tieback system discussed w/OCFD but no approval. Will be changing from rescue to recovery mode and will use long arm excavator to remove debris from under the Mother Slab—most of the remaining victims are now known to be in that location
	1000	Boldt has finished collar connections & shear bolts at pipe, lateral bracing cols. F20 & 22 to E 18, 20, & 22 Brief CA-TF4 (Orange Co.) struct. spec—final TF to be oncoming
	1100	Give all clear to remove conc. slabs on north face between F20 & 22 w/communication to smart level watcher
	1200	Work w/front OPS regarding lifts of slabs. Decided to leave Christmas tree slabs at col. F18 since no victims are left there. Work will concentrate on area between cols. 22 & 24 back to under Mother Slab—start use of excavator
	1300	Checked Mother Slab cables hourly + check Smart Levels when lifts occur; 11 continue monitor w/theodolite (East Tower + Mother Slab). Low winds
	1400	Brief John Osteraas—replacement lead IST engineer
	1630	Lead IST engineer Dave leaves site for home—Heartfelt thanks to all personnel—BE SAFE. Engineer John will replace Engineer Dave
	1800	Typ Co-ordination meeting—no major issues
	late P.M.	Continue monitor and aid w/lifting large slabs
1 MAY 95	0700	Typ. coordination meeting—no major issues bracing completed in Pit
	0800	Continue monitoring of East Tower + cols. F22 & 20 + Mother Slab. ALL STABLE

		Work shut down part time due to weather overnight
	0900	Inspection of bracing in Pit area showed that small correction is needed
	1000	Monitoring continued—no movement
		Jay Flintco was shown bracing that needed correction—correction completed and col. F22 now more secure—more debris may be removed
	1200	Granite panel no. 3 at east end of 6th floor removed to eliminate falling hazard
	1300	Mother Slab is being monitored continuously—no sign of movement or cracking. Compared picture from 23 Apr 95 to current cracking—no significant change in slab—continue with current monitor schedule
	1400	Continue monitoring of all controls & Mother Slab. No movement
		Assist OPS as requested w/slab, col. & beam picks
	1800	Typ. coordination meeting—no major issues
		Discontinue night operations for remainder of recovery operation
2–5 MAY 95	0700– 1800	Operations continued without significant structural issue and operation was officially ended at 2345 on 5 May 95
		Two victims were left in the very complicated pile of rubble at the base of col. F22 due to safety considerations. The area was clearly marked and they were removed the day after the remainder of the building was demolished by the demolition contractor.

REFERENCES

Abbott, J. F., et al. (1995). "Measurements to determine the detonability of 10 gallon ammonium nitrate mixtures." *Report no. FaAA-SF-R-95-07-05*, Failure Analysis Associates, Inc., Menlo Park, California, July (proprietary).

Arson and explosives incidents report. (1994). Department of the Treasury, Bureau of Alcohol, Tobacco and Firearms.

Conrath, E., and Walton, B. (1995). "Terror strikes Oklahoma City." *Security Engineering Update*, U.S. Army Corps of Engineers, Omaha District, Aug.

Council directive of 15 July 1980 on the approximation of the laws of the member states relating to straight ammonium nitrate fertilizers in high nitrogen content. (1980). *Title 80/876/EEC*, Commission of the European Communities, July 15.

Deposition of Henry C. Gibbons. (1995). Criminal Complaint, United States of America v. Timothy James McVeigh, United States District Court, Western District of Oklahoma, Case no. M-95-98-H, Apr. 21.

"Details and detailing of concrete reinforcement." (1986). *Technical Report No. ACI 315-80*, revised American Concrete Institute, Detroit, Mich.

Gorman, C., (1995). "The bomb lurking in the garden shed." *Time*, (May), 54.

Harvey, R. N. (1995). "The Port Authority of New York and New Jersey's organizational strategy for recovering the World Trade Center after the February 26, 1993, terrorist bombing." *Cost Engineering*, Jan.

Holzer, T. L., et al. (1995). "Interpretation of seismographs of the April 19, 1995, Oklahoma City, Oklahoma, bombing." U.S. Geological Survey, Menlo Park, Calif. (to be submitted to *Science*, Sept. 27).

King, A., and Bauer, A. (1977). "A review of accidents with ammonium nitrate, prepared for the Canadian Fertilizer Institute and Contributing Bodies." the Department of Mining Engineering, Queen's University, Kingston, Ontario, June 29.

Kupperman, R., and Trent D. (1979). *Terrorism*. Hoover Institution Press, Stanford University, Stanford, Calif.

Mitchell, D., and Cook, W. (1984). "Preventing progressive collapse of slab structures." *Journal of Structural Engineering*, ASCE, 110 (7).

Partin, B.K. (1995). "Bomb damage analysis of Alfred P. Murrah federal building, Oklahoma City, Oklahoma." Alexandria, Va., July 13.

Porter, S. J. (1968). Method of desensitizing fertilizer grade ammonium nitrate and the product obtained. Patent 3, 366, 468, U.S. Patent Office, Jan. 30.

BIBLIOGRAPHY

The design of structures against the effects of explosions has traditionally been the domain of the military. Although the methodologies are similar for both military and civilian construction, the design objectives are different. For instance, in military applications the primary objective is to preserve the operational capability of the facility. In civilian applications, the loss of life is a primary issue. Also, civilian structures need to be more accessible to the public, which raises architectural issues that are absent in military construction.

There are few standards and guidelines available at this time for civilian protective design. Engineering judgment and the needs/limitations of the client are typically used to arrive at a design. The purpose of this bibliography is to provide a resource for engineers charged with the responsibility of designing protective structures who need practical design references to enable them to make informed design decisions.

Military Technical Manuals

The U.S. government has sponsored a vast amount of research in the area of explosion effects. Although the primary focus of this literature is on military applications, much of it is relevant to civilian protective construction.

There are numerous documents published by the U.S. Army, U.S. Navy, and U.S. Air Force in the area of explosion effects. Although some are not available to the general public because of the sensitive nature of the topic, many of them are. Government publications can be ordered from the National Technical Information Service and/or the Defense Technical Information Center.

Three of the most recent publications are:

Dobbs, N., et al. (1987). *Structures to resist the effects of accidental explosions* (6 vols.). Special publication ARLCD-SP-84001. U.S. Army Research, Development, and Engineering Center, Armament Engineering Directorate, Picatinny Arsenal, N.J., Dec.

Drake, J. L., et al. (1989). *Protective construction design manual* (8 vols.). Final report ESL-TR-87-57. Air Force Engineering & Services Center, Engineering & Services Laboratory, Tyndall Air Force Base, Fla., Nov.

Fundamentals of protective design for conventional weapons. (1986). TM 5-855-1, HQ Dept. of the Army, Washington, D.C., Nov. 3.

Security engineering. (1994). Army TM 5-853, Air Force AFMAN 32-1071 (3 vols.). Depts. of the Army and Air Force, May 12.

Earlier publications are listed below. Most contain the same or similar information to the documents referenced above.

Ammunition explosive safety standards. (1984). DOD 6055-9-STD, Dept. of Defense, Washington, D.C., July.

Baker, W. E., et al. (1980). *A manual for the prediction of blast and fragment loading on structures.* Report no. DOE/TIC-11268, U.S. Army Engineer Division, Huntsville, Ala., Nov.

Crawford, R. E., et al. (1974). *The Air Force manual for design and analysis of hardened structures.* Final report AFWL-TR-74-102, Air Force Weapons Laboratory, Air Force Systems Command, Kirtland Air Force Base, N.M., Oct.

Crawford, R. E., et al. (1971). *Protection from nonnuclear weapons.* Technical report no. AFWL-TR-70-127, Air Force Weapons Laboratory, Kirtland Air Force Base, N.M., Feb.

Designing facilities to resist nuclear weapon effects. (1979). Technical manual no. TM 5-858-3, HQ Dept. of the Army, Washington, D.C., Mar.

Newmark, N. M., and Haltiwanger, J. D. (1962). *Air Force design manual: principles and practices for design of hardened structures.* Report no. AFSWC-TDR-62-138, Air Force Special Weapons Center, Air Force Systems Command, Kirtland Air Force Base, N.M., Dec.

Protective construction concepts. (1968). HQ U.S. Air Force, Directorate of Civil Engineering, Washington, D.C., Nov.

Suppressive shields: Structural design and analysis handbook. (1977). Report no. HNDM-1110-1-2, U.S. Army Corps of Engineers, Huntsville Division, Nov.

Although most of the information in the handbooks published before 1984 is still valid, some of it has been updated. Several of the more useful references in this category are:

Kingery, C. N., and Bulmash, G. (1984). *Airblast parameters from TNT spherical air burst and hemispherical surface burst.* Technical report ARBL-TR-02555, Ballistic Research Laboratory, Aberdeen Proving Ground, Md., Apr. (contains updated blast parameter data.)

Woodson, S. C., and Kiger, S. (1988). "Stirrup requirements, for blast resistant slabs." *Journal of Structural Engineering*, ASCE, 114(9) (contains results of explosive tests that indicate that stirrups may be as effective as lacing in some situations).

Nonmilitary Publications

Most of the handbooks listed below provide design information similar to that found in the government handbooks.

Baker, W. E., et al. (1983). *Fundamental Studies in Engineering 5: Explosion Hazards and Evaluation.* Elsevier Scientific Publishing Co., New York, N.Y.

Design of structures to resist nuclear weapons effects. (1985). ASCE—Manuals and Reports on Engineering Practice No. 42, Revised Ed., ASCE, New York, N.Y.

Goschy, B. (1989). *Design of Buildings to Withstand Abnormal Loading.* Butterworths, London.

Kinney, G. F., and Graham, K. J. (1985). *Explosive Shocks in Air*, 2nd Ed., Springer-Verlag, New York, N.Y.

The texts listed below provide a theoretical background to structural dynamics, which enables the designer to apply dynamic analysis methods to a wide range of problems.

Biggs, J. M. (1964). *Introduction to Structural Dynamics*, McGraw Hill Book Co. New York, N.Y.

Blevins, R. D. (1979). *Formulas for Natural Frequency and Mode Shape.* Van Nostrand Reinhold Co., New York, N.Y.

Craig, R. R., Jr. (1981). *Structural Dynamics: an Introduction to Computer Methods.* John Wiley & Sons. New York, N.Y.

Harris, C. M., and Crede, C. E. (1976). *Shock and Vibration Handbook,* 2nd Ed., McGraw Hill Book Co., New York, N.Y.

Norris, C. H., et al. (1959). *Structural Design for Dynamic Loads.* McGraw Hill Book Co., New York, N.Y.

Timoshenko, S. P. (1937). *Vibration Problems in Engineering,* 2nd Ed., Van Nostrand Reinhold Co., New York, N.Y.

Progressive Collapse

Progressive collapse is an important issue in the design of buildings against explosion effects, particularly for structural precast concrete construction. Consult the following texts:

Breen, J. E., and Siess, C. P. (1979). "Progressive collapse—symposium summary." *ACI Journal,* (Sept.), 997–1005.

Burnett, E. F. P. (1975). *The avoidance of progressive collapse: regulatory approaches to the problem.* NBS-GCR 75048, National Bureau of Standards, Washington, D.C., Oct.

Ellingwood, B., and Leyendecker, E. V. (1978). "Approaches for design against progressive collapse." *Journal of the Structural Division,* ASCE, 104(3), 413–423.

Leyendecker, E. V., and Ellingwood, B. R. (1977). *Design methods for reducing the risk of progressive collapse in buildings.* NBS Building Science Series 98, National Bureau of Standards, Washington, D.C. Apr.

Minimum design loads for buildings and other structures. (1995). ASCE 7-95, revision of ANSI/ASCE 7-93, ASCE, New York, N.Y.

Speyer, I. J. (1976). "Considerations for the design of precast concrete bearing wall buildings to withstand abnormal loads." *PCI Journal,* (Mar.–Apr.) 18–51.

INDEX